Robotic Computing
on FPGAs

Synthesis Lectures on Computer Architecture

Editor
Natalie Enright Jerger, *University of Toronto*

Editor Emerita
Margaret Martonosi, *Princeton University*

Founding Editor Emeritus
Mark D. Hill, *University of Wisconsin, Madison*

Synthesis Lectures on Computer Architecture publishes 50- to 100-page books on topics pertaining to the science and art of designing, analyzing, selecting, and interconnecting hardware components to create computers that meet functional, performance, and cost goals. The scope will largely follow the purview of premier computer architecture conferences, such as ISCA, HPCA, MICRO, and ASPLOS.

Quantum Computer System: Research for Noisy Intermediate-Scale Quantum Computers
Yongshan Ding and Frederic T. Chong
2020

A Primer on Memory Consistency and Cache Coherence, Second Edition
Vijay Nagarajan, Daniel J. Sorin, Mark D. Hill, and David Wood
2020

Innovations in the Memory System
Rajeev Balasubramonian
2019

Cache Replacement Policies
Akanksha Jain and Calvin Lin
2019

The Datacenter as a Computer: Designing Warehouse-Scale Machines, Third Edition
Luiz André Barroso, Urs Hölzle, and Parthasarathy Ranganathan
2018

Principles of Secure Processor Architecture Design
Jakub Szefer
2018

General-Purpose Graphics Processor Architectures
Tor M. Aamodt, Wilson Wai Lun Fung, and Timothy G. Rogers
2018

Compiling Algorithms for Heterogenous Systems
Steven Bell, Jing Pu, James Hegarty, and Mark Horowitz
2018

Architectural and Operating System Support for Virtual Memory
Abhishek Bhattacharjee and Daniel Lustig
2017

Deep Learning for Computer Architects
Brandon Reagen, Robert Adolf, Paul Whatmough, Gu-Yeon Wei, and David Brooks
2017

On-Chip Networks, Second Edition
Natalie Enright Jerger, Tushar Krishna, and Li-Shiuan Peh
2017

Robotic Computing on FPGAs

Shaoshan Liu, Zishen Wan, Bo Yu, and Yu Wang

ISBN: 978-3-031-00643-2   paperback
ISBN: 978-3-031-01771-1   ebook
ISBN: 978-3-031-00068-3   hardcover

DOI 10.1007/978-3-031-01771-1

A Publication in the Springer series
SYNTHESIS LECTURES ON COMPUTER ARCHITECTURE

Lecture #56
Series Editor: Natalie Enright Jerger, *University of Toronto*
Editor Emerita: Margaret Martonosi, *Princeton University*
Founding Editor Emeritus: Mark D. Hill, *University of Wisconsin, Madison*
Series ISSN
Print 1935-3235   Electronic 1935-3243

Robotic Computing on FPGAs

Shaoshan Liu
PerceptIn

Zishen Wan
Georgia Institute of Technology

Bo Yu
PerceptIn

Yu Wang
Tsinghua University

SYNTHESIS LECTURES ON COMPUTER ARCHITECTURE #56

ABSTRACT

This book provides a thorough overview of the state-of-the-art field-programmable gate array (FPGA)-based robotic computing accelerator designs and summarizes their adopted optimized techniques. This book consists of ten chapters, delving into the details of how FPGAs have been utilized in robotic perception, localization, planning, and multi-robot collaboration tasks. In addition to individual robotic tasks, this book provides detailed descriptions of how FPGAs have been used in robotic products, including commercial autonomous vehicles and space exploration robots.

KEYWORDS

robotics, FPGAs, autonomous machines, perception, localization, planning, control, space exploration, deep learning

Contents

Preface

In this book, we provide a thorough overview of the state-of-the-art FPGA-based robotic computing accelerator designs and summarize their adopted optimized techniques. The authors combined have over 40 years of research experiences of utilizing FPGAs in robotic applications, both in academic research and commercial deployments. For instance, the authors have demonstrated that, by co-designing both the software and hardware, FPGAs can achieve more than 10× better performance and energy efficiency compared to the CPU and GPU implementations. The authors have also pioneered the utilization of the partial reconfiguration methodology in FPGA implementations to further improve the design flexibility and reduce the overhead. In addition, the authors have successfully developed and shipped commercial robotic products powered by FPGAs and the authors demonstrate that FPGAs have excellent potential and are promising candidates for robotic computing acceleration due to its high reliability, adaptability, and power efficiency.

The authors believe that FPGAs are the best compute substrate for robotic applications for several reasons. First, robotic algorithms are still evolving rapidly, and thus any ASIC-based accelerators will be months or even years behind the state-of-the-art algorithms. On the other hand, FPGAs can be dynamically updated as needed. Second, robotic workloads are highly diverse, thus it is difficult for any ASIC-based robotic computing accelerator to reach economies of scale in the near future. On the other hand, FPGAs are a cost effective and energy-effective alternative before one type of accelerator reaches economies of scale. Third, compared to systems on a chip (SoCs) that have reached economies of scale, e.g., mobile SoCs, FPGAs deliver a significant performance advantage. Fourth, partial reconfiguration allows multiple robotic workloads to time-share an FPGA, thus allowing one chip to serve multiple applications, leading to overall cost and energy reduction.

Specifically, FPGAs require little power and are often built into small systems with less memory. They have the ability of massively parallel computations and to make use of the properties of perception (e.g., stereo matching), localization (e.g., simultaneous localization and mapping (SLAM)), and planning (e.g., graph search) kernels to remove additional logic so as to simplify the end-to-end system implementation. Taking into account hardware characteristics, several algorithms are proposed which can be run in a hardware-friendly way and achieve similar software performance. Therefore, FPGAs are possible to meet real-time requirements while achieving high energy efficiency compared to central processing units (CPUs) and graphics processing units (GPUs). In addition, unlike the application-specific integrated circuit (ASIC) counterparts, FPGA technologies provide the flexibility of on-site programming and re-programming without going through re-fabrication with a modified design. Partial Reconfiguration (PR) takes this flexibility one step further, allowing the modification of an operating FPGA design by loading a partial configuration file. Using PR, part of the FPGA can be reconfigured at runtime without compromising the integrity of the applications running on those parts of the device that are not being reconfigured. As a result, PR can allow different robotic applications to time-share part of an FPGA, leading to energy and performance efficiency, and making FPGA a suitable computing platform for dynamic and complex robotic workloads. Due to the advantages over other compute substrates, FPGAs have been successfully utilized in commercial autonomous vehicles as well as in space robotic applications, for FPGAs offer unprecedented flexibility and significantly reduced the design cycle and development cost.

This book consists of ten chapters, providing a thorough overview of how FPGAs have been utilized in robotic perception, localization, planning, and multi-robot collaboration tasks. In addition to individual robotic tasks, we provide detailed descriptions of how FPGAs have been used in robotic products, including commercial autonomous vehicles and space exploration robots.

Shaoshan Liu
June 2021

CHAPTER 1

Introduction and Overview

The last decade has seen significant progress in the development of robotics, spanning from algorithms, mechanics to hardware platforms. Various robotic systems, like manipulators, legged robots, unmanned aerial vehicles, and self-driving cars have been designed for search and rescue [1, 2], exploration [3, 4], package delivery [5], entertainment [6, 7], and more applications and scenarios. These robots are on the rise of demonstrating their full potential. Take drones, a type of aerial robot, as an example. The number of drones has grown by 2.83x between 2015 and 2019 based on the U.S. Federal Aviation Administration (FAA) report [8]. The registered number reached 1.32 million in 2019, and the FFA expects this number will grow to 1.59 billion by 2024.

However, robotic systems are very complex [9–12]. They tightly integrate many technologies and algorithms, including sensing, perception, mapping, localization, decision making, control, etc. This complexity poses many challenges for the design of robotic edge computing systems [13, 14]. On the one hand, robotic systems need to process an enormous amount of data in real-time. The incoming data often comes from multiple sensors, which is highly heterogeneous and requires accurate spatial and temporal synchronization and pre-processing [15]. However, the robotic system usually has limited on-board resources, such as memory storage, bandwidth, and compute capabilities, making it hard to meet the real-time requirements. On the other hand, the current state-of-the-art robotic system usually has strict power constraints on the edge that cannot support the amount of computation required for performing tasks, such as 3D sensing, localization, navigation, and path planning. Therefore, the computation and storage complexity, as well as real-time and power constraints of the robotic system, hinder its wide application in latency-critical or power-limited scenarios [16].

Therefore, it is essential to choose a proper compute platform for robotic systems. CPUs and GPUs are two widely used commercial compute platforms. CPUs are designed to handle a wide range of tasks quickly and are often used to develop novel algorithms. A typical CPU can achieve 10–100 GFLOPS with below 1 GOP/J power efficiency [17]. In contrast, GPUs are designed with thousands of processor cores running simultaneously, which enable massive parallelism. A typical GPU can perform up to 10 TOPS performance and become a good candidate for high-performance scenarios. Recently, benefiting in part from the better accessibility provided by CUDA/OpenCL, GPUs have been predominantly used in many robotic applications. However, conventional CPUs and GPUs usually consume 10–100 W of power, which are orders of magnitude higher than what is available on the resource-limited robotic system.

Besides CPUs and GPUs, FPGAs are attracting attention and becoming compute substrate candidates to achieve energy-efficient robotic task processing. FPGAs require low power and are often built into small systems with less memory. They have the ability to process massively parallel computations and to make use of the properties of perception (e.g., stereo matching), localization (e.g., SLAM), and planning (e.g., graph search) kernels to remove additional logic and simplify the implementation. Taking into account hardware characteristics, researchers and engineers have proposed several algorithms that can be run in a hardware-friendly way and achieve similar software performance. Therefore, FPGAs are possible to meet real-time requirements while achieving high energy efficiency compared to CPUs and GPUs.

Unlike the ASIC counterparts, FPGAs provide the flexibility of on-site programming and re-programming without going through re-fabrication with a modified design. Partial Reconfiguration (PR) takes this flexibility one step further, allowing the modification of an operating FPGA design by loading a partial configuration file. By using PR, part of the FPGA can be reconfigured at runtime without compromising the integrity of the applications running on the parts of the device that are not being reconfigured. As a result, PR can allow different robotic applications to time-share part of an FPGA, leading to energy and performance efficiency, and making FPGA a suitable computing platform for dynamic and complex robotic workloads.

Note that robotics is not one technology but rather an integration of many technologies. As shown in Fig. 1.1, the stack of the robotic system consists of three major components: application workloads, including sensing, perception, localization, motion planning, and control; a software edge subsystem, including operating system and runtime layer; and computing hardware, including both micro-controllers and companion computers [16, 18, 19].

We focus on the robotic application workloads in this chapter. The application subsystem contains multiple algorithms that are used by the robot to extract meaningful information from raw sensor data to understand the environment and dynamically make decisions about its actions.

1.1   SENSING

The sensing stage is responsible for extracting meaningful information from the sensor raw data. To enable intelligent actions and improve reliability, the robot platform usually supports a wide range of sensors. The number and type of sensors are heavily dependent on the specifications of the workload and the capability of the onboard compute platform. The sensors can include the following:

Cameras. Cameras are usually used for object recognition and object tracking, such as lane detection in autonomous vehicles and obstacle detection in drones, etc. RGB-D camera can also be utilized to determine object distances and positions. Take autonomous vehicles as an example, the current system usually mounts eight or more 1080p cameras around the vehicle to detect, recognize and track objects in different directions, which can greatly improve safety.

Usually, these cameras run at 60 Hz, which will process about multiple gigabytes of raw data per second when combined.

GNSS/IMU. The global navigation satellite system (GNSS) and inertial measurement unit (IMU) system help the robot localize itself by reporting both inertial updates and an estimate of the global location at a high rate. Different robots have different requirements for localization sensing. For instance, 10 Hz may be enough for a low-speed mobile robot, but high-speed autonomous vehicles usually demand 30 Hz or higher for localization, and high-speed drones may need 100 Hz or more for localization, thus we are facing a wide spectrum of sensing speeds. Fortunately, different sensors have their advantages and drawbacks. GNSS can enable fairly accurate localization, while it runs at only 10 Hz, thus unable to provide real-time updates. By contrast, both accelerometer and gyroscope in IMU can run at 100–200 Hz, which can satisfy the real-time requirement. However, IMU suffers bias wandering over time or perturbation by some thermo-mechanical noise, which may lead to an accuracy degradation in the position estimates. By combining GNSS and IMU, we can get accurate and real-time updates for robots.

LiDAR. Light detection and ranging (LiDAR) is used for evaluating distance by illuminating the obstacles with laser light and measuring the reflection time. These pulses, along with other recorded data, can generate precise and three-dimensional information about the surrounding characteristics. LiDAR plays an important role in localization, obstacle detection, and avoidance. As indicated in [20], the choice of sensors dictates the algorithm and hardware design. Take autonomous driving as an instance, almost all the autonomous vehicle companies use LiDAR at the core of their technologies. Examples include Uber, Waymo, and Baidu. PerceptIn and Tesla are among the very few that do not use LiDAR and, instead, rely on cameras and vision-based systems. In particular, PerceptIn's data demonstrated that for the low-speed autonomous driving scenario, LiDAR processing is slower than camera-based vision processing, but increases the power consumption and cost.

Radar and Sonar. The Radio Detection and Ranging (Radar) and Sound Navigation and Ranging (Sonar) system is used to determine the distance and speed to a certain object, which usually serves as the last line of defense to avoid obstacles. Take autonomous vehicles as an example, a danger of collision may occur when near obstacles are detected, then the vehicle will apply brakes or turn to avoid obstacles. Compared to LiDAR, the Radar and Sonar system is cheaper and smaller, and their raw data is usually fed to the control processor directly without going through the main compute pipeline, which can be used to implement some urgent functions as swerving or applying the brakes.

One key problem we have observed with commercial CPUs, GPUs, or mobile SoCs is the lack of built-in multi-sensor processing supports, hence most of the multi-sensor processing has to be done in software, which could lead to problems such as time synchronization. On the other hand, FPGAs provide a rich sensor interface and enable most time-critical sensor

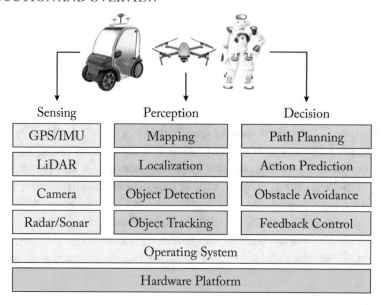

Figure 1.1: The stack of the robotic system.

processing tasks to be done in hardware [21]. In Chapter 2, we introduce FPGA technologies, especially how FPGAs provide rich I/O blocks, which can be configured for heterogeneous sensor processing.

1.2   PERCEPTION

The sensor data is then fed into the perception layer to sense the static and dynamic objects as well as build a reliable and detailed representation of the robot's environment by using computer vision techniques (including deep learning).

The perception layer is responsible for object detection, segmentation, and tracking. There are obstacles, lane dividers, and other objects to detect. Traditionally, a detection pipeline starts with image pre-processing, followed by a region of interest detector, and finally a classifier that outputs detected objects. In 2005, Dalal and Triggs [22] proposed an algorithm based on the histogram of orientation (HOG) and support vector machine (SVM) to model both the appearance and shape of the object under various condition. The goal of segmentation is to give the robot a structured understanding of its environment. Semantic segmentation is usually formulated as a graph labeling problem with vertices of the graph being pixels or super-pixels. Inference algorithms on graphical models such as conditional random field (CRF) [23, 24] are used. The goal of tracking is to estimate the trajectory of moving obstacles. Tracking can be formulated as a sequential Bayesian filtering problem by recursively running the prediction step and correction step. Tracking can also be formulated by tracking-by-detection handling with

Markovian decision process (MDP) [25], where an object detector is applied to consecutive frames and detected objects are linked across frames.

In recent years, deep neural networks (DNNs), also known as deep learning, have greatly affected the field of computer vision and made significant progress in solving robot perception problems. Most state-of-the-art algorithms now apply one type of neural network based on convolution operation. Fast R-CNN [26], Faster R-CNN [27], SSD [28], YOLO [29], and YOLO9000 [30] were used to get much better speed and accuracy in object detection. Most CNN-based semantic segmentation work is based on Fully Convolutional Networks (FCNs) [31], and there are some recent work in spatial pyramid pooling network [32] and pyramid scene parsing network (PSPNet) [33] to combine global image-level information with the locally extracted feature. By using auxiliary natural images, a stacked autoencoder model can be trained offline to learn generic image features and then applied for online object tracking [34].

In Chapter 3, we review the state-of-the-art neural network accelerator designs and demonstrate that with software-hardware co-design, FPGAs can achieve more than 10 times better speed and energy efficiency than the state-of-the-art GPUs. This verifies that FPGAs are a promising candidate for neural network acceleration. In Chapter 4, we review various stereo vision algorithms in the robotic perception and their FPGA accelerator designs. We demonstrate that with careful algorithm-hardware co-design, FPGAs can achieve two orders of magnitude of higher energy efficiency and performance than the state-of-the-art GPUs and CPUs.

1.3   LOCALIZATION

The localization layer is responsible for aggregating data from various sensors to locate the robot in the environment model.

GNSS/IMU system is used for localization. The GNSS consist of several satellite systems, such as GPS, Galileo, and BeiDou, which can provide accurate localization results but with a slow update rate. In comparison, IMU can provide a fast update with less accurate rotation and acceleration results. A mathematical filter, such as Kalman Filter, can be used to combine the advantages of the two and minimize the localization error and latency. However, this sole system has some problems, such as the signal may bounce off obstacles, introduce more noise, and fail to work in closed environments.

LiDAR and High-Definition (HD) maps are used for localization. LiDAR can generate point clouds and provide a shape description of the environment, while it is hard to differentiate individual points. HD map has a higher resolution compared to digital maps and makes the route familiar to the robot, where the key is to fuse different sensor information to minimize the errors in each grid cell. Once the HD map is built, a particle filter method can be applied to localize the robot in real-time correlated with LiDAR measurement. However, the LiDAR performance may be severely affected by weather conditions (e.g., rain, snow) and bring localization error.

Cameras are used for localization as well. The pipeline of vision-based localization is simplified as follows: (1) by triangulating stereo image pairs, a disparity map is obtained and used to derive depth information for each point; (2) by matching salient features between successive stereo image frames in order to establish correlations between feature points in different frames, the motion between the past two frames is estimated; and (3) by comparing the salient features against those in the known map, the current position of the robot is derived [35].

Apart from these techniques, sensor fusion strategy is also often utilized to combine multiple sensors for localization, which can improve the reliability and robustness of robot [36, 37].

In Chapter 5, we introduce a general-purpose localization framework that integrates key primitives in existing algorithms along with its implementation in FPGA. The FPGA-based localization framework retains high accuracy of individual algorithms, simplifies the software stack, and provides a desirable acceleration target.

1.4   PLANNING AND CONTROL

The planning and control layer is responsible for generating trajectory plans and passing the control commands based on the original and destination of the robot. Broadly, prediction and routing modules are also included here, where their outputs are fed into downstream planning and control layers as input. The prediction module is responsible for predicting the future behavior of surrounding objects identified by the perception layer. The routing module can be a lane-level routing based on lane segmentation of the HD maps for autonomous vehicles.

Planning and control layers usually include behavioral decision, motion planning, and feedback control. The mission of the behavioral decision module is to make effective and safe decisions by leveraging all various input data sources. Bayesian models are becoming more and more popular and have been applied in recent works [38, 39]. Among the Bayesian models, the Markov Decision Process (MDP) and Partially Observable Markov Decision Process (POMDP) are the widely applied methods in modeling robot behavior. The task of motion planning is to generate a trajectory and send it to the feedback control for execution. The planned trajectory is usually specified and represented as a sequence of planned trajectory points, and each of these points contains attributes like location, time, speed, etc. Low-dimensional motion planning problems can be solved with grid-based algorithms (such as Dijkstra [40] or A* [41]) or geometric algorithms. High-dimensional motion planning problems can be dealt with sampling-based algorithms, such as Rapidly exploring Random Tree (RRT) [42] and Probabilistic Roadmap (PRM) [43], which can avoid the problem of local minima. Reward-based algorithms, such as the Markov decision process (MDP), can also generate the optimal path by maximizing cumulative future rewards. The goal of feedback control is to track the difference between the actual pose and the pose on the predefined trajectory by continuous feedback. The most typical and widely used algorithm in robot feedback control is PID.

While optimization-based approaches enjoy mainstream appeal in solving motion planning and control problems, learning-based approaches [44–48] are becoming increasingly popular with recent developments in artificial intelligence. Learning-based methods, such as reinforcement learning, can naturally make full use of historical data and iteratively interact with the environment through actions to deal with complex scenarios. Some model the behavioral level decisions via reinforcement learning [46, 48], while other approaches directly work on motion planning trajectory output or even direct feedback control signals [45]. Q-learning [49], Actor-Critic learning [50], and policy gradient [43] are some popular algorithms in reinforcement learning.

In Chapter 6, we introduce the motion planning modules of the robotics system, and compare several FPGA and ASIC accelerator designs in motion planning to analyze intrinsic design trade-offs. We demonstrate that with careful algorithm-hardware co-design, FPGAs can achieve three orders of magnitude than CPUs and two orders of magnitude than GPUs with much lower power consumption. This demonstrates that FPGAs can be a promising candidate for accelerating motion planning kernels.

1.5   FPGAS IN ROBOTIC APPLICATIONS

Besides accelerating the basic modules in the robotic computing stack, FPGAs have been utilized in different robotic applications. In Chapter 7, we explore how FPGAs can be utilized in multi-robot exploration tasks. Specifically, we present an FPGA-based interruptible CNN accelerator and a deployment framework for multi-robot exploration.

In Chapter 8, we provide a retrospective summary of PerceptIn's efforts on developing on-vehicle computing systems for autonomous vehicles, especially how FPGAs are utilized to accelerate critical tasks in a full autonomous driving stack. For instance, localization is accelerated on an FPGA while depth estimation and object detection are accelerated by a GPU. This case study has demonstrated that FPGAs are capable of playing a crucial role in autonomous driving, and exploiting accelerator-level parallelism while taking into account constraints arising in different contexts could significantly improve on-vehicle processing.

In Chapter 9, we explore how FPGAs have been utilized in space robotic applications in the past two decades. The properties of FPGAs make them good onboard processors for space missions, ones that have high reliability, adaptability, processing power, and energy efficiency. FPGAs may help us close the two-decade performance gap between commercial processors and space-grade ASICs when it comes to powering space exploration robots.

1.6   THE DEEP PROCESSING PIPELINE

Different from other computing workloads, autonomous machines have a very deep processing pipeline with strong dependencies between different stages and a strict time-bound associated with each stage [51]. For instance, Fig. 1.2 presents an overview of the processing pipeline of an autonomous driving system. Starting from the left side, the system consumes raw sensing data from mmWave radars, LiDARs, cameras, and GNSS/IMUs, and each sensor produces raw data at a different frequency. The cameras capture images at 30 FPS and feed the raw data to the *2D Perception module*, the LiDARs capture point clouds at 10 FPS and feed the raw data to the

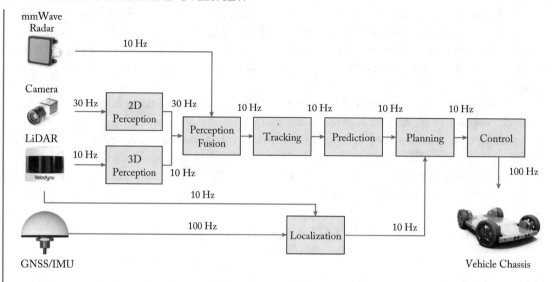

Figure 1.2: The processing pipeline of autonomous vehicles.

3D Perception module as well as the *Localization module*, the GNSS/IMUs generate positional updates at 100 Hz and feed the raw data to the *Localization module*, the mmWave radars detect obstacles at 10 FPS and feed the raw data to the *Perception Fusion module*.

Next, the results of 2D and 3D Perception Modules are fed into the *Perception Fusion module* at 30 Hz and 10 Hz, respectively, to create a comprehensive perception list of all detected objects. The perception list is then fed into the *Tracking module* at 10 Hz to create a tracking list of all detected objects. The tracking list is then fed into the *Prediction module* at 10 Hz to create a prediction list of all objects. After that, both the prediction results and the localization results are fed into the *Planning module* at 10 Hz to generate a navigation plan. The navigation plan is then fed into the *Control module* at 10 Hz to generate control commands, which are finally sent to the autonomous vehicle for execution at 100 Hz.

Hence, for each 10 ms, the autonomous vehicle needs to generate a control command to maneuver the vehicle. If any upstream module, such as the *Perception module*, misses the deadline to generate an output, the *Control module* still has to generate a command before the deadline. This could lead to disastrous results as the autonomous vehicle is essentially driving blindly without the perception output.

The key challenge is to design a system to minimize the end-to-end latency of the deep processing pipeline within energy and cost constraints, and with minimum latency variation. In this book, we demonstrate that FPGAs can be utilized in different modules in this long processing pipeline to minimize latency, reduce latency variation, and achieve energy efficiency.

1.7   SUMMARY

The authors believe that FPGAs are the indispensable compute substrate for robotic applications for several reasons.

- First, robotic algorithms are still evolving rapidly. Thus, any ASIC-based accelerators will be months or even years behind the state-of-the-art algorithms; on the other hand, FPGAs can be dynamically updated as needed.

- Second, robotic workloads are highly diverse. Thus, it is difficult for any ASIC-based robotic computing accelerator to reach economies of scale in the near future; on the other hand, FPGAs are a cost-effective and energy-effective alternative before one type of accelerator reaches economies of scale.

- Third, compared to SoCs that have reached economies of scale, e.g., mobile SoCs and FPGAs deliver a significant performance advantage.

- Fourth, partial reconfiguration allows multiple robotic workloads to time-share an FPGA, thus allowing one chip to serve multiple applications, leading to overall cost and energy reduction.

Specifically, FPGAs require little power and are often built into small systems with less memory. They have the ability to parallel computations massively and make use of the properties of perception (e.g., stereo matching), localization (e.g., SLAM), and planning (e.g., graph search) kernels to remove additional logic and simplify the implementation. Taking into account hardware characteristics, several algorithms are proposed which can be run in a hardware-friendly way and achieve similar software performance. Therefore, FPGAs are possible to meet real-time requirements while achieving high energy efficiency compared to CPUs and GPUs. Unlike the ASIC counterparts, FPGAs provide the flexibility of on-site programming and re-programming without going through re-fabrication with a modified design. PR takes this flexibility one step further, allowing the modification of an operating FPGA design by loading a partial configuration file. Using PR, part of the FPGA can be reconfigured at runtime without compromising the integrity of the applications running on those parts of the device that are not being reconfigured. As a result, PR can allow different robotic applications to time-share part of an FPGA, leading to energy and performance efficiency, and making FPGA a suitable computing platform for dynamic and complex robotic workloads.

Due to the advantages over other compute substrates, FPGAs have been successfully utilized in commercial autonomous vehicles. Particularly, over the past four years, PerceptIn has built and commercialized autonomous vehicles for micromobility, and PerceptIn's products have been deployed in China, the U.S., Japan, and Switzerland. In this book, we provide a real-world case study on how PerceptIn developed its computing system by relying heavily on FPGAs, which perform not only heterogeneous sensor synchronizations but also the acceleration of software components on the critical path. In addition, FPGAs are used heavily in space robotic applications, for FPGAs offered unprecedented flexibility and significantly reduced the design cycle and development cost.

CHAPTER 2

FPGA Technologies

Before we delve into utilizing FPGAs for accelerating robotic workloads, in this chapter we first provide the background of FPGA technologies so that readers without prior knowledge can grasp the basic understanding of what an FPGA is and how an FPGA works. We also introduce partial reconfiguration, a technique that exploits the flexibility of FPGAs and one that is extremely useful for various robotic workloads to time-share an FPGA so as to minimize energy consumption and resource utilization. In addition, we explore existing techniques that enable the robot operating system (ROS), an essential infrastructure for robotic computing, to run directly on FPGAs.

2.1   AN INTRODUCTION TO FPGA TECHNOLOGIES

In the 1980s, FPGAs emerged as a result of increasing integration in electronics. Before the use of FPGAs, glue-logic designs were based on individual boards with fixed components interconnected via a shared standard bus, which has various drawbacks, such as hindrance of high volume data processing and higher susceptibility to radiation-induced errors, in addition to inflexibility.

In detail, FPGAs are semiconductor devices that are based around a matrix of configurable logic blocks (CLBs) connected via programmable interconnects. FPGAs can be reprogrammed to desired application or functionality requirements after manufacturing. This feature distinguishes FPGAs from Application-Specific Integrated Circuits (ASICs), which are custom manufactured for specific design tasks.

Note that ASICs and FPGAs have different value propositions, and they must be carefully evaluated before choosing any one over the other. While FPGAs used to be selected for lower-speed/complexity/volume designs in the past, today's FPGAs easily push the 500 MHz performance barrier. With unprecedented logic density increases and a host of other features, such as embedded processors, DSP blocks, clocking, and high-speed serial at ever lower price points, FPGAs are a compelling proposition for almost any type of design.

Modern FPGAs are with massive reconfigurable logic and memory, which let engineers build dedicated hardware with superb power and performance efficiency. Especially, FPGAs are attracting attention from the robotic community and becoming an energy-efficient platform for robotic computing. Unlike ASIC counterparts, FPGA technology provides the flexibility of on-site programming and re-programming without going through re-fabrication with a modified design, due to its underlying reconfigurable fabrics.

2.1.1   TYPES OF FPGAS

FPGAs can be categorized by the type of their programmable interconnection switches: antifuse, SRAM, and Flash. Each of the three technologies comes with trade-offs.

- **Antifuse FPGAs** are non-volatile and have a minimal delay due to routing, resulting in a faster speed and lower power consumption. The drawback is evident as they have a relatively more complicated fabrication process and are only one-time programmable.

- **SRAM-based FPGAs** are field reprogrammable and use the standard fabrication process that foundries put in significant effort in optimizing, resulting in a faster rate of performance increase. However, based on SRAM, these FPGAs are volatile and may not hold configuration if a power glitch occurs. Also, they have more substantial routing delays, require more power, and have a higher susceptibility to bit errors. Note that SRAM-based FPGAs are the most popular compute substrates in space applications.

- **Flash-based FPGAs** are non-volatile and reprogrammable, and also have low power consumption and route delay. The major drawback is that runtime reconfiguration is not recommended for flash-based FPGAs due to the potentially destructive results if radiation effects occur during the reconfiguration process [52]. Also, the stability of stored charge on the floating gate is of concern: it is a function including factors such as operating temperature, the electric fields that might disturb the charge. As a result, flash-based FPGAs are not as frequently used in space missions [53].

2.1.2   FPGA ARCHITECTURE

In this subsection, we introduce the basic components in FPGA architecture in the hope of providing basic background knowledge to readers with limited prior knowledge on FPGA technologies. For a detailed and thorough explanation, interested authors can refer to [54].

As shown in Fig. 2.1, a basic FPGA design usually contains the following components.

- **Configurable Logic Blocks (CLBs)** are the basic repeating logic resources on an FPGA. When linked together by the programmable routing blocks, CLBs can execute complex logic functions, implement memory functions, and synchronize code on the FPGA. CLBs contain smaller components, including flip-flops (FFs), look-up tables (LUTs), and multiplexers (MUX). An FF is the smallest storage resource on the FPGA. Each FF in a CLB is a binary register used to save logic states between clock cycles on an FPGA circuit. An LUT stores a predefined list of outputs for every combination of inputs. LUTs provide a fast way to retrieve the output of a logic operation because possible results are stored and then referenced rather than calculated. A MUX is a circuit that selects between two or more inputs and then returns the selected input. Any logic can be implemented using the combination of FFs, LUTs, and MUX.

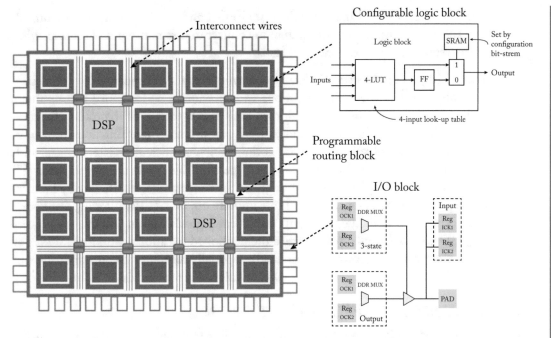

Figure 2.1: Overview of FPGA architecture.

- **Programmable Routing Blocks (PRBs)** provide programmability for connectivity among a pool of CLBs. The interconnection network contains configurable switch matrices and connection blocks that can be programmed to form the demanded connection. PRBs can be divided into Connection Blocks (CBs) and a matrix of Switch Boxes (SBs), namely, Switch Matrix (SM). CBs are responsible to provide a connection between CLBs input/output pins to the adjacent routing channels. SBs are placed at the intersection points of vertical and horizontal routing channels. Routing a net from a CLB source to a CLB target necessitates passing through multiple interconnect wires and SBs, in which an entering signal from a certain side can connect to any of the other three directions based on the SM topology.

- **I/O Blocks (IOBs)** are used to bridge signals onto the chip and send them back off again. An IOB consists of an input buffer and an output buffer with three-state and open-collector output controls. Typically, there are pull-up resistors on the outputs and sometimes pull-down resistors that can be used to terminate signals and buses without requiring discrete resistors external to the chip. The polarity of the output can usually be programmed for active high or active low output. There are typical flip-flops on outputs so that clocked signals can be output directly to the pins without encountering significant delay, more easily meeting the setup time requirement for external devices.

Since there are many IOBs available on an FPGA and these IOBs are programmable, we can easily design a compute system to connect to different types of sensors, which are extremely useful in robotic workloads.

- **Digital Signal Processors (DSPs)** have been optimized to implement various common digital signal processing functions with maximum performance and minimum logic resource utilization. In addition to multipliers, each DSP block has functions that are frequently required in typical DSP algorithms. These functions usually include pre-adders, adders, subtractors, accumulators, coefficient register storage, and a summation unit. With these rich features, the DSP blocks in the Stratix series FPGAs are ideal for applications with high-performance and computationally intensive signal processing functions, such as finite impulse response (FIR) filtering, fast Fourier transforms (FFTs), digital up/down conversion, high-definition (HD) video processing, HD CODECs, etc. Besides the aforementioned traditional workloads, DSPs are also extremely useful for robotic workloads, especially computer vision workloads, providing high-performance and low-power solutions for robotic vision front ends [55].

2.1.3   COMMERCIAL APPLICATIONS OF FPGAS

Due to their programmable nature, FPGAs are an ideal fit for many different markets such as the following.

- **Aerospace & Defense** – Radiation-tolerant FPGAs along with the intellectual property for image processing, waveform generation, and partial reconfiguration for Software-Defined Radios, especially for space and defense applications.

- **ASIC Prototyping** – ASIC prototyping with FPGAs enables fast and accurate SoC system modeling and verification of embedded software.

- **Automotive** – FPGAs enable automotive silicon and IP solutions for gateway and driver assistance systems, as well as comfort, convenience, and in-vehicle infotainment.

- **Consumer Electronics** – FPGAs provide cost-effective solutions enabling next-generation, full-featured consumer applications, such as converged handsets, digital flat panel displays, information appliances, home networking, and residential set top boxes.

- **Data Center** – FPGAs have been utilized heavily for high-bandwidth, low-latency servers, networking, and storage applications to bring higher value into cloud deployments.

- **High-Performance Computing and Data Storage** – FPGAs have been utilized widely for Network Attached Storage (NAS), Storage Area Network (SAN), servers, and storage appliances.

- **Industrial** – FPGAs have been utilized in targeted design platforms for Industrial, Scientific, and Medical (ISM) enable higher degrees of flexibility, faster time-to-market, and lower overall non-recurring engineering costs (NRE) for a wide range of applications such as industrial imaging and surveillance, industrial automation, and medical imaging equipment.

- **Medical** – For diagnostic, monitoring, and therapy applications, FPGAs have been used to meet a range of processing, display, and I/O interface requirements.

- **Security** – FPGAs offer solutions that meet the evolving needs of security applications, from access control to surveillance and safety systems.

- **Video & Image Processing** – FPGAs have been utilized in targeted design platforms to enable higher degrees of flexibility, faster time-to-market, and lower overall non-recurring engineering costs (NRE) for a wide range of video and imaging applications.

- **Wired Communications** – FPGAs have been utilized to develop end-to-end solutions for the Reprogrammable Networking Linecard Packet Processing, Framer/MAC, serial backplanes, and more.

- **Wireless Communications** – FPGAs have been utilized to develop RF, base band, connectivity, transport, and networking solutions for wireless equipment, addressing standards such as WCDMA, HSDPA, WiMAX, and others.

In the rest of this book, we explore robotic computing, an emerging and potentially a killer application for FPGAs. With FPGAs, we can develop low-power, high-performance, cost-effective, and flexible compute systems for various robotic workloads. Due to the advantages provided by FPGAs, we expect that robotic applications will be a major demand driver for FPGAs in the near future.

2.2   PARTIAL RECONFIGURATION

Unlike the ASIC counterparts, FPGAs provide the flexibility of on-site programming and re-programming without going through re-fabrication with a modified design. PR takes this flexibility one step further, allowing the modification of an operating FPGA design by loading a PR file. Using PR, part of the FPGA can be reconfigured at runtime without compromising the integrity of the applications running on those parts of the device that are not being reconfigured. As a result, PR can allow different robotic applications to time-share part of an FPGA, leading to energy and performance efficiency, and making FPGAs suitable computing platforms for dynamic and complex robotic workloads.

2.2.1 WHAT IS PARTIAL RECONFIGURATION?

The obvious benefit of using reconfigurable devices, such as FPGAs, is that the functionality that a device has now can be changed and updated at some time in the future. As additional functionality is available or design improvements are made available, the FPGA can be completely reprogrammed with new logic. PR takes this capability one step further by allowing designers to change the logic within a part of an FPGA without disrupting the entire system. This allows designers to divide their system into modules, each comprised of one block of logic and, without disrupting the whole system and stopping the flow of data, the users can update the functionality within one block.

Runtime partial reconfiguration (RPR) is a special feature offered by many FPGAs that allows designers to reconfigure certain portions of the FPGA during runtime without influencing other parts of the design. This feature allows the hardware to be adaptive to a changing environment. First, it allows optimized hardware implementation to accelerate computation. Second, it allows efficient use of chip area such that different hardware modules can be swapped in/out of the chip at runtime. Last, it may allow leakage and clock distribution power saving by unloading hardware modules that are not active.

RPR is extremely useful for robotic applications, as a mobile robot might encounter very different environments as it navigates, and it might require different perception, localization, or planning algorithms for these different environments. For instance, while a mobile robot is in an indoor environment, it is likely to use an indoor map for localization, but when it travels outdoor, it might choose to use GPS and visual-inertial odometry for localization. Keeping multiple hardware accelerators for different tasks is not only costly but also energy inefficient. RPR provides a perfect solution for this problem. As shown in Fig. 2.2, an FPGA is divided into three partitions for the three basic functions, one for perception, one for localization, and one for planning. Then for each function, there are three algorithms ready, one for each environment. Each of these algorithms is converted to a bit file and ready for RPR when needed. For instance, when a robot navigates to a new environment and decides that a new perception algorithm is needed, it can load the target bit file and sends it to the internal configuration access port (ICAP) to reconfigure the perception partition.

One major challenge of RPR for robotic computing is the configuration speed, as most robotic tasks have strong real-time constraints, and to maintain the performance of the robot, the reconfiguration process has to finish within a very tight time bound. In addition, the reconfiguration process incurs performance and power overheads. By maximizing the configuration speed, these overheads can be minimized as well.

2.2.2 HOW TO USE PARTIAL RECONFIGURATION?

PR allows the modification of an operating FPGA design by loading a PR file, or a bit file through ICAP [56]. Using PR, after a full bit file configures the FPGA, partial bit files can also be downloaded to modify reconfigurable regions in the FPGA without compromising the

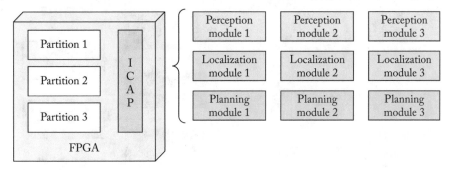

Figure 2.2: An example of partial reconfiguration for robotic applications.

integrity of the applications running on those parts of the device that are not being reconfigured. RPR allows a limited, predefined portion of an FPGA to be reconfigured while the rest of the device continues to operate, and this feature is especially valuable where devices operate in a mission-critical environment that cannot be disrupted while some subsystems are being redefined.

In an SRAM-based FPGA, all user-programmable features are controlled by memory cells that are volatile and must be configured on power-up. These memory cells are known as the configuration memory, and they define the look-up table (LUT) equations, signal routing, input/output block (IOB) voltage standards, and all other aspects of the design. In order to program the configuration memory, instructions for the configuration control logic and data for the configuration memory are provided in the form of a bitstream, which is delivered to the device through the JTAG, SelectMAP, serial, or ICAP configuration interface. An FPGA can be partially reconfigured using a partial bitstream. A designer can use such a partial bitstream to change the structure of one part of an FPGA design as the rest of the device continues to operate.

RPR is useful for systems with multiple functions that can time-share the same FPGA device resources. In such systems, one section of the FPGA continues to operate, while other sections of the FPGA are disabled and reconfigured to provide new functionality. This is analogous to the situation where a microprocessor manages context switching between software processes. In the case of PR of an FPGA, however, it is the hardware instead of the software that is being switched.

RPR provides an advantage over multiple full bitstreams in applications that require continuous operation, which would not be possible during full reconfiguration. One example is a mobile robot that switches the perception module while keeping the localization module and planning module intact when moving from a dark environment to a bright environment. With RPR, the system can maintain the localization and planning modules while the perception module within the FPGA is changed on the fly.

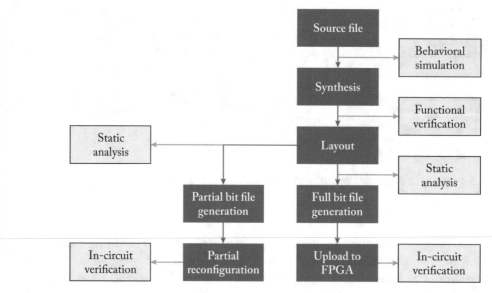

Figure 2.3: FPGA regular and partial reconfiguration design flow.

Xilinx has provided the PR feature in their high-end FPGAs, the Virtex series, in limited access BETA since the late 1990s. More recently it is a production feature supported by their tools and across their devices since the release of ISE 12. The support for this feature continues to improve in the more recent release of ISE 13. Altera has promised this feature for their new high-end devices, but this has not yet materialized. PR of FPGAs is a compelling design concept for general purpose reconfigurable systems for its flexibility and extensibility.

Using the Xilinx tool chain, designers can go through the regular synthesis flow to generate a single bitstream for programming the FPGA. This considers the device as a single atomic entity. As opposed to the general synthesis flow, the PR flow physically divides the FPGA device into regions. One region is called the "static region," which is the portion of the device that is programmed at startup and never changes. Another region is the "PR region," which is the portion of the device that will be reconfigured dynamically, potentially multiple times and with different designs. It is possible to have multiple PR regions, but we will consider only the simplest case here. The PR flow generates at least two bitstreams, one for the static and one for the PR region. Most likely, there will be multiple PR bitstreams, one for each design that can be dynamically loaded.

As shown in Fig. 2.3, the first step in implementing a system using the PR design flow is the same as the regular design, which is to synthesize the netlists from the HDL sources that will be used in the implementation and layout process. Note that the process requires separate netlists for the static (top-level) designs and the PR partitions. A netlist must be generated for each implementation of the PR partition used in the design. If the system design has multiple

PR partitions, then it will require a netlist for each implementation of each PR partition, even if the logic is the same in multiple locations. Then once a netlist is done, we need to work on the layout for each design to make sure that the netlist fits into the dedicated partition, and we need to make sure that there are enough resources available for the design in each partition. Once the implementation is done, we can then generate the bit file for each partition. At runtime, we can dynamically swap different designs to a partition for the robot to adapt to the changing environment. For more details on how to use PR on FPGAs, please refer to [57].

2.2.3 ACHIEVING HIGH PERFORMANCE

A major performance bottleneck for PR is the configuration overhead, which determines the usefulness of PR. If PR is done fast enough, we can use this feature to enable mobile robots to swap hardware components at runtime. If PR cannot be done fast enough, we can only use this feature to perform offline hardware updates.

To address this problem, in [58], the authors propose a combination of two techniques to minimize the overhead. First, the authors design and implement fully streaming DMA engines to saturate the configuration throughput. Second, the authors exploit a simple form of data redundancy to compress the configuration bitstreams, and implement an intelligent internal configuration access port (ICAP) controller to perform decompression at runtime. This design achieves an effective configuration data transfer throughput of up to 1.2 GB/s, which well surpasses the theoretical upper bound of the data transfer throughput, 400 MB/s. Specifically, the proposed fully streaming DMA engines reduce the configuration time from the range of seconds to the range of milliseconds, a more than 1000-fold improvement. In addition, the proposed compression scheme achieves up to a 75% reduction in bitstream size and results in a decompression circuit with negligible hardware overhead.

Figure 2.4 shows the architecture of the fast PR engine, which consists of:

- a direct memory access (DMA) engine to establish a direct transfer link between the external SRAM, where the configuration files are stored, and the ICAP;

- a streaming engine implemented with a FIFO queue to buffer data between the consumer and the producer to eliminate the handshake between the producer and the consumer for each data transfer; and

- turn on the burst mode for ICAP thus it can fetch four words instead of one word at a time.

We will explain this design in greater details in the following sections.

Problems with the Out-of-Box PR Engine Design
Without the fast PR engine, in the out-of-box design, the ICAP Controller contains only the ICAP and the ICAP FSM, and the SRAM Controller only contains the SRAM Bridge and

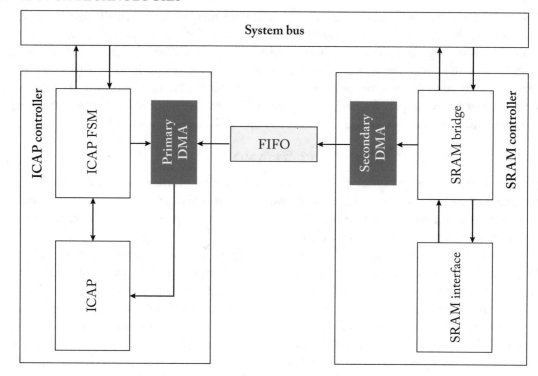

Figure 2.4: Fast partial reconfiguration engine.

the SRAM Interface. Hence, there is no direct memory access between SRAM and ICAP, and all configuration data transfers are done in software. In this way, the pipeline issues one read instruction to fetch a configuration word from SRAM, and then issues a write instruction to send the word to ICAP; instructions are also fetched from SRAM, and this process repeats until the transfer process completes. This scheme is highly inefficient because the transfer of one word requires tens of cycles, and the ICAP transfer throughput of this design is only 318 KB/s, whereas on the product specification, the ideal ICAP throughput is 400 MB/s. Hence the out-of-box design throughput is 1000 times worse than the ideal design.

Configuration Time is a Pure Function of the Bitstream Size?

Theoretically, the ICAP throughput can reach 400 MB/s, but this is achievable only if the configuration time is a pure function of bitstream file size. In order to find out whether this theoretical throughput is achievable, the authors of [58] performed experiments to configure different regions of the FPGA chip, to repeatedly writing NOPs to ICAP, and to stress the configuration circuit by repeatedly configuring one region. During all these tests, we found out that ICAP always ran at full speed such that it was able to consume four bytes of configuration data per cycle, regardless of the semantics of the configuration data. This confirms that configuration time is a pure function of the size of the bitstream file.

Adding the Primary-Secondary DMA Engines

To improve PR throughput, we first can simply implement a pair of primary-secondary DMA engines. The primary DMA engine resides in the ICAP controller and interfaces with the ICAP FSM, the ICAP, as well as the secondary DMA engine. The secondary DMA engine resides in the SRAM Controller, and it interfaces with the SRAM Bridge and the primary DMA engine. When a DMA operation starts, the primary DMA engine receives the starting address as well as the size of the DMA operation. Then it starts sending control signals (read-enable, address, etc.) to the secondary DMA engine, which then forwards the signals to the SRAM Bridge. After the data is fetched, the secondary DMA engine sends the data back to the primary DMA engine. Then, the primary DMA engine decrements the size counter, increments the address, and repeats the process to fetch the next word. Compared to the out-of-box design, simply adding the DMA engines avoids the involvement of the pipeline in the data transfer process and it significantly increases the PR throughput to 50 MB/s, a 160-fold improvement.

Adding a FIFO between the DMA Engines

To further improve the PR throughput, we can modify the primary-secondary DMA engines by adding a FIFO between the two DMA engines. In this version of the design, when DMA operation starts, instead of sending control signals to the secondary DMA engine, the primary DMA engine forwards the starting address and the size of the DMA operation to the secondary DMA engine, then it waits for the data to become available in the FIFO. Once data becomes available in the FIFO, the primary DMA engine reads the data and decrements its size counter. When the counter hits zero, the DMA operation completes. On the other side, upon receiving the starting address and size of the DMA operation, the secondary DMA engine starts sending control signals to the SRAM Bridge to fetch data one word at a time. Then once the secondary DMA engine receives data from the SRAM Bridge, it writes the word into the FIFO, decrements its size counter, and increments its address register to fetch the next word. In this design, only data is transferred between the primary and secondary DMA engines, and all control operations to SRAM are handled in the secondary DMA. This greatly simplifies the handshaking between the ICAP Controller and the SRAM Controller, and it leads to a 100 MB/s ICAP throughput, an additional two-fold improvement.

Adding Burst Mode to Provide Fully Streaming

The SRAM on most FPGA boards usually provides burst read mode such that we can read four words at a time instead of one. Burst mode reads are available on most DDR memories as well. There is an ADVLD signal to the SRAM device. During a read, if this signal is set, then a new address is loaded into the device. Otherwise, the device will output a burst of up to four words, one word per cycle. Therefore, if we can set the ADVLD signal every four cycles, each time we increment the address by four words, and given that the synchronization between control signals and data fetches is correct, then we are able to stream data from SRAM to the ICAP. We implement two independent state machines in the secondary DMA engine. One state machine sends control signals as well as the addresses to the SRAM in a continuous manner, such that in every four cycles, the address is incremented by four words (16 bytes) and sent to the SRAM device. The other state machine simply waits for the data to become ready at the beginning, and then in each cycle, it receives one word from the SRAM and streams the word to the FIFO until the DMA operation completes. Similarly, the primary DMA engine waits for data to become available in the FIFO, and then in each cycle, it reads one word from the FIFO and streams the word to the ICAP until the DMA operation completes. This fully streaming DMA design leads to an ICAP throughput that exceeds 395 MB/s, which is very close to the ideal 400 MB/s throughputs.

Energy Efficiency

In [59], the authors indicate that the polarity of the FPGA hardware structures may significantly impact leakage power consumption. Based on this observation, the authors of [60] tried to find out whether FPGAs utilize this property such that when the blank bitstream is loaded to wipe out an accelerator, the circuit is set to a state to minimize the leakage power consumption. In order to achieve this, the authors implemented eight PR regions on an FPGA chip, with each region occupying a configuration frame. These eight PR regions did not consume any dynamic power, as the authors purposely gated off the clock to these regions. Then the authors used the blank bitstream files to wipe out each of these regions and observed the chip power consumption behavior. The results indicated that for every four configuration frames that we applied the blank bitstream on, the chip power consumption dropped by a constant amount. This study confirms that PR indeed leads to static power reduction and suggests that FPGAs may have utilized the polarity property to minimize leakage power.

In addition, the authors of [60] studied whether PR can be used as an effective energy reduction technique in reconfigurable computing systems. To approach this problem, the authors first identified the analytical models that capture the necessary conditions for energy reduction under different system configurations. The models show that increasing the configuration throughput is a general and effective way to minimize the PR energy overhead. Therefore, the authors designed and implemented a fully streaming DMA engine that nearly saturates the configuration throughput.

The findings provide answers to the three questions: first, although we pay extra power to use an accelerator, depending on the accelerator's ability to accelerate the program execution, it will result in actual energy reduction. The experimental results in [60] demonstrate that due to its low power overhead and excellent ability of acceleration, having an acceleration extension can lead to both program speedup and system energy reduction. Second, it is worthwhile to use PR to reduce chip energy consumption if the energy reduction can make up for the energy overhead incurred during the reconfiguration process; and the key to minimize the energy overhead during the reconfiguration process is to maximize the configuration speed. The experimental results in [60] confirm that enabling PR is a highly effective energy reduction technique. Finally, clock gating is an effective technique in reducing energy consumption due to its negligible overhead; however, it reduces only dynamic power whereas PR reduces both dynamic and static power. Therefore, PR can lead to a larger energy reduction than clock gating, provided the extra energy saving on static power elimination can make up for the energy overhead incurred during the reconfiguration process.

Although the conventional wisdom is that PR is only useful if the accelerator would not be used for a very long period of time, the experimental results in [60] indicate that with the high configuration throughput delivered by the fast PR engine, PR can outperform clock gating in energy reduction even if the accelerator inactive time is in the millisecond range. In summary, based on the results from [58] and [60], we can conclude that PR is an effective technique for improving both performance and energy efficiency, and it is the key feature that makes FPGAs a highly attractive choice for dynamic robotic computing workloads.

2.2.4   REAL-WORLD CASE STUDY

Following the design presented in [60], PerceptIn has demonstrated in their commercial product that RPR is useful for robotic computing, especially computing for autonomous vehicles, because many on-vehicle tasks usually have multiple versions where each is used in a particular scenario [20]. For instance, in PerceptIn's design, the localization algorithm relies on salient features; features in keyframes are extracted by a feature extraction algorithm (based on ORB features [61]), whereas features in non-key frames are tracked from previous frames (using optical flow [62]); the latter executes in 10 ms, 50% faster than the former. Spatially sharing the FPGA is not only area-inefficient but also power-inefficient as the unused portion of the FPGA consumes non-trivial static power. In order to temporally share the FPGA and "hot-swap" different algorithms, PerceptIn developed a Partial Reconfiguration Engine (PRE) that dynamically reconfigures part of the FPGA at runtime. The PRE achieves a 400 MB/s reconfiguration throughput (i.e., bitstream programming rate). Both the feature extraction and tracking bitstreams are less than 4 MB. Thus, the reconfiguration delay is less than 1 ms.

2.3   ROBOT OPERATING SYSTEM (ROS) ON FPGAS

As demonstrated in the previous chapter, autonomous vehicles and robots demand complex information processing such as SLAM (Simultaneous Localization and Mapping), deep learning, and many other tasks. FPGAs are promising in accelerating these applications with high energy efficiency. However, utilizing FPGAs for robotic workloads is challenging due to the high development costs and lack of talents who can understand both FPGAs and robotics. One way to address this challenge is to directly support ROS on FPGAs as ROS already provides the basic infrastructure for supporting efficient robotic computing. Hence, in this section we explore the state-of-the-art supports for ROS to run on FPGAs.

2.3.1   ROBOT OPERATING SYSTEM (ROS)

Before delving into supports for running ROS on FPGAs, we first understand the importance of ROS in robotic applications. ROS is an open-source, meta-operating system for autonomous machines and robots. It provides the essential operating system services, including hardware abstraction, low-level device control, implementation of commonly used functionality, message-passing between processes, and package management. ROS also provides tools and libraries for obtaining, building, writing, and running code across multiple computers. The primary goal of ROS is to support code reuse in robotics research and development. In essence, ROS is a distributed framework of processes that enables executables to be individually designed and loosely coupled at runtime. These processes can be grouped into Packages and Stacks, which can be easily shared and distributed. ROS also supports a federated system of code Repositories that enable collaboration to be distributed as well. This design, from the file system level to the community level, enables independent decisions about development and implementation, but all can be brought together with ROS infrastructure tools [63].

The core objectives of the ROS framework include the following.

- Thin: ROS is designed to be as thin as possible so that code written for ROS can be used with other robot software frameworks.

- ROS-agnostic libraries: the preferred development model is to write ROS-agnostic libraries with clean functional interfaces.

- Language independence: the ROS framework is easy to implement in any modern programming language. The ROS development team has already implemented it in Python, C++, and Lisp, and we have experimental libraries in Java and Lua.

- Easy testing: ROS has a built-in unit/integration test framework called rostest that makes it easy to bring up and tear down test fixtures.

- Scaling: ROS is appropriate for large runtime systems and large development processes.

The Computation Graph is the peer-to-peer network of ROS processes that are processing data together. The basic Computation Graph concepts of ROS are nodes, Master, Parameter Server, messages, services, topics, and bags, all of which provide data to the Graph in different ways.

- Nodes: nodes are processes that perform computation. ROS is designed to be modular at a fine-grained scale; a robot control system usually comprises many nodes. Take autonomous vehicles as an example, one node controls a laser range-finder, one node controls the wheel motors, one node performs localization, one node performs path planning, one node provides a graphical view of the system, and so on. A ROS node is written with the use of a ROS client library, such as roscpp or rospy.

- Master: the ROS Master provides name registration and lookup to the rest of the Computation Graph. Without the Master, nodes would not be able to find each other, exchange messages, or invoke services.

- Parameter Server: the parameter server allows data to be stored by key in a central location. It is currently part of the Master.

- Messages: nodes communicate with each other by passing messages. A message is simply a data structure, comprising typed fields. Standard primitive types (integer, floating-point, boolean, etc.) are supported, as are arrays of primitive types. Messages can include arbitrarily nested structures and arrays (much like C structs).

- Topics: messages are routed via a transport system with publish-subscribe semantics. A node sends out a message by publishing it to a given topic. The topic is a name that is used to identify the content of the message. A node that is interested in a certain kind of data will subscribe to the appropriate topic. There may be multiple concurrent publishers and subscribers for a single topic, and a single node may publish and subscribe to multiple topics. In general, publishers and subscribers are not aware of each others' existence. The idea is to decouple the production of information from its consumption. Logically, one can think of a topic as a strongly typed message bus. Each bus has a name, and anyone can connect to the bus to send or receive messages as long as they are the right type.

- Services: the publish-subscribe model is a very flexible communication paradigm, but its many-to-many, one-way transport is not appropriate for request-reply interactions, which are often required in a distributed system. Request-reply is done via services, which are defined by a pair of message structures: one for the request and one for the reply. A providing node offers a service under a name and a client uses the service by sending the request message and awaiting the reply. ROS client libraries generally present this interaction to the programmer as if it were a remote procedure call.

- Bags: bags are a format for saving and playing back ROS message data. Bags are an important mechanism for storing data, such as sensor data, that can be difficult to collect but is necessary for developing and testing algorithms.

The ROS Master acts as a name service in the ROS Computation Graph. It stores topics and services registration information for ROS nodes. Nodes communicate with the Master to report their registration information. As these nodes communicate with the Master, they can receive information about other registered nodes and make connections as appropriate. The Master

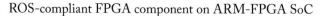

ROS-compliant FPGA component on ARM-FPGA SoC

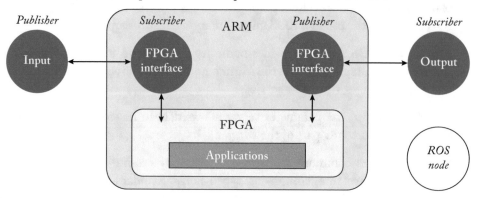

Figure 2.5: ROS-compliant FPGAs.

will also make callbacks to these nodes when this registration information changes, which allows nodes to dynamically create connections as new nodes are run.

Nodes connect to other nodes directly; the Master only provides lookup information, much like a domain name service (DNS) server. Nodes that subscribe to a topic will request connections from nodes that publish that topic and will establish that connection over an agreed-upon connection protocol. This architecture allows for decoupled operations, where the names are the primary means by which larger and more complex systems can be built. Names have a very important role in ROS: nodes, topics, services, and parameters all have names. Every ROS client library supports command-line remapping of names, which means a compiled program can be reconfigured at runtime to operate in a different Computation Graph topology.

2.3.2   ROS-COMPLIANT FPGAS

In order to integrate FPGAs into a ROS-based system, a ROS-compliant FPGA component has been proposed [64, 65]. Integration of an FPGA into a robotic system requires equivalent functionality to replace a software ROS component with a ROS-compliant FPGA component. Therefore, each ROS message type and data format used in the ROS-compliant FPGA component must be the same as that of the software ROS component. The ROS-compliant FPGA component aims to improve its processing performance while satisfying the requirements.

Figure 2.5 shows the architecture of the ROS-compliant FPGA component model. Each ROS-compliant FPGA component must implement the following four functions: Encapsulation of FPGA circuits, Interface between ROS software and FPGA circuits, Subscribe interface from a topic, and Publish interface to a topic. The ARM core is responsible for communicating with and offloading workloads to the FPGA, whereas the FPGA part performs actual workload acceleration. Note that there are two roles of software in the component. First, an interface process for input that subscribes to a topic to receive input data. The software component, which runs on the ARM core, is responsible for formatting the data suitable for the FPGA processing and sends the formatted data to the FPGA. Second, an interface process for output receives processing results from the FPGA. The software component, which runs on the ARM core, is responsible for reformatting the results suitable for the ROS system and publishes them to a topic. Such a structure can realize a robot system in which software and hardware cooperate.

Note that the difference of ROS-compliant FPGA component from a ROS node written in pure software is that processing contains hardware processing of an FPGA. Integration of ROS-compliant FPGA component into a ROS system only requires connections to ROS nodes through Publish/Subscribe messaging in ordinary ROS development style. The ROS-compliant FPGA component provides easy integration of an FPGA by wrapping it with software.

To evaluate this design, the authors of [65] have implemented a hardwired image labeling application on a ROS-compliant FPGA component on Xilinx Zynq-7020, and verifying that this design performs 26 times faster than that of software with the ARM processor, and even 2.3 times faster than that of an Intel PC. Moreover, the end-to-end latency of the component is 1.7 times faster than that of processing with pure software. Therefore, the authors verify that the ROS-compliant FPGA component achieves remarkable performance improvement, maintaining high development productivity by cooperative processing of hardware and software. However, this also comes with a problem, as the authors found out that the communication of ROS nodes is a major bottleneck of the execution time in the ROS-compliant FPGA component.

2.3.3   OPTIMIZING COMMUNICATION LATENCY FOR THE ROS-COMPLIANT FPGAS

As indicated in the previous subsection, large communication latency between ROS components is a severe problem and has been the bottleneck of offloading computing to FPGAs. The authors in [66] aim to reduce the latency by implementing Publish/Subscribe messaging of ROS as hardware. Based on the results of network packets analysis in the ROS system, the authors propose a method of implementing a hardware ROS-compliant FPGA Component, which is done by separating the registration part (XMLRPC) and data communication part (TCPROS) of the Publish/Subscribe messaging.

To study ROS performance, the authors have compared the communication latency of (1) PC-PC and (2) PC-ARM SoC. Two computer nodes are connected with each other through a Gigabit Ethernet. The communication latency in (2) PC-ARM SoC environment is about four times larger than (1) PC-PC. Therefore, the performance in embedded processor environments, such as ARM processors, should be improved. Hence, the challenge for ROS-compliant FPGA components is to reduce the large overhead in communication latency. If communication latency is reduced, the ROS-compliant FPGA component can be used as an accelerator for processing in robotic applications/systems.

In order to implement Publish/Subscribe messaging of ROS as hardware, the authors analyzed network packets that flowed in Publish/Subscribe messaging in the ROS system of ordinary software. The authors utilized WireShark for network packet analysis [67] with the basic ROS setup of one master, one publisher, and one subscriber node.

- STEP (1): the Publisher and Subscriber nodes register their nodes and topic information to the Master node. The registration is done by calling methods like registerPublisher, hasParam, and so on, using XMLRPC [68].

- STEP (2): the Master node notifies topic information to the Subscriber nodes by calling publisherUpdate (XMLRPC).

- STEP (3): the Subscriber node sends a connection request to the Publisher node by using requestTopic (XMLRPC).

- STEP (4): the Publisher node returns IP address and port number, TCP connection information for data communication, as a response to the requestTopic (XMLRPC).

- STEP (5): the Subscriber node establishes a TCP connection by using the information and sends connection header to the TCP connection. Connection header contains important metadata about a connection being established, including typing and routing information, using TCPROS [69].

- STEP (6): if it is a successful connection, the Publisher node sends connection header (TCPROS).

- STEP (7): data transmission repeats. This data is written with little endian and header information (4 bytes) is added to the data (TCPROS).

After this analysis, the authors found out that network packets that flowed in Publish/Subscribe messaging in the ROS system can be categorized into two parts, that is, the registration part and the data transmission part. The registration part uses XMLRPC (STEPS (1)–(4)), while the data transmission part uses TCPROS (STEPS (5)–(7)), which is almost raw data of TCP communication with very small overhead. In addition, once data transmission (STEP (7)) starts, only data transmission repeats without STEPS (1)–(6).

Based on the network packet analysis, the authors modified the server ports, such that those used in XMLRPC and TCPROS are assigned differently. In addition, a client TCP/IP connection of XMLRPC for the Master node is necessary for the Publisher node. For the Subscriber node, two client TCP/IP connections of XMLRPC and one client connection of TCPROS are necessary. Therefore, two or three TCP ports are necessary to implement Publish/Subscribe messaging. It is a problem to implement ROS nodes using the hardware TCP/IP stack.

To optimize the communication performance on ROS-compliant FPGAs, the authors proposed hardware publication and subscription services. Conventionally, publication or subscription of topics was done by software in ROS. By implementing these nodes as hardwired circuits, direct communication between the ROS nodes and the FPGA becomes not only possible but also highly efficient. In order to implement the hardware ROS nodes, the authors designed the Subscriber hardware and the Publisher hardware separately: the Subscriber hardware is responsible to subscribe to a topic of another ROS node and to receive ROS messages from the topic; whereas the Publisher hardware is responsible to publish ROS messages to a topic of another ROS node. With this hardware-based design, the evaluation results indicate that the latency of the Hardware ROS-compliant FPGA component can be cut to half, from 1.0 ms to 0.5 ms, thus effectively improving the communication between the FPGA accelerator and other software-based ROS nodes.

2.4   SUMMARY

In this chapter, we have provided a general introduction to FPGA technologies, especially runtime partial reconfiguration, which allows multiple robotic workloads to time-share an FPGA at runtime. We also have introduced existing research on enabling ROS on FPGAs, which provides infrastructure supports for various robotic workloads to run directly on FPGAs. However, the ecosystem of robotic computing on FPGAs is still in its infancy. For instance, due to the lack of high-level synthesis tools for robotic accelerator design, accelerating robotic workloads, or part of a robotic workload, on FPGAs still require extensive manual efforts. To make the matter worse, most robotic engineers do not have sufficient FPGA background to develop an FPGA-based accelerator, whereas few FPGA engineers possess sufficient robotic background to fully understand a robotic system. Hence, to fully exploit the benefits of FPGAs, advanced design automation tools are imperative to bridge this knowledge gap.

CHAPTER 3

Perception on FPGAs – Deep Learning

Cameras are widely used in intelligent robot systems because of their lightweight and rich information for perception. Cameras can be used to complete a variety of basic tasks of intelligent robots, such as visual odometry (VO), place recognition, object detection, and recognition. With the development of convolutional neural networks (CNNs), we can reconstruct the depth and pose with the absolute scale directly from a monocular camera, making monocular VO more robust and efficient. And monocular VO methods, like Depth-VO-Feat [70], make robot systems much easier to deploy than stereo ones. Furthermore, although there are previous works to design accelerators for robot applications, such as ESLAM [71], the accelerators can only be used for specific applications with poor scalability.

In recent years, CNN has made great improvements on the place recognition for robotic perception. The accuracy of the place recognition code from another CNN-based method, GeM [72], is about 20% better than the handcrafted method, rootSIFT [73]. CNN is a general framework, which can be applied to a variety of robotic applications. With the help of CNN, the robots can also detect and distinguish objects from input images. In summary, CNNs greatly enhance robots' ability in localization, place recognition, and many other perception tasks.

CNNs have become the core component in various kinds of robots. However, since neural networks (NNs) are computationally intensive, deep learning models are often the performance bottleneck in robots. In this chapter, we delve into utilizing FPGAs to accelerate neural networks in various robotic workloads.

Specifically, neural networks are widely adopted in regions like image, speech, and video recognition. What's more, deep learning has made significant progress in solving robotic perception problems. But the high computation and storage complexity of neural network inference poses great difficulty in its application. CPUs are hard to offer enough computational capacity. GPUs are the first choice for the neural network process because of their high computational capacity and easy-to-use development frameworks but suffer from energy inefficiency.

On the other hand, with specifically designed hardware, FPGAs are a potential candidate to surpass GPUs in performance and energy efficiency. Various FPGA-based accelerators have been proposed with software and hardware optimization techniques to achieve high performance and energy efficiency. In this chapter, we give an overview of previous work on neural network inference accelerators based on FPGAs and summarize the main techniques used. An

Table 3.1: Performance and resource utilization of state-of-the-art neural network accelerator designs

	AlexNet[74]	VGG19[78]	ResNet152[81]	MobileNet[79]	ShuffleNet[80]
Year	2012	2014	2016	2017	2017
# Param	60M	144M	57M	4.2M	2.36M
# Operation	1.4G	39G	22.6G	1.1G	0.27G
Top-1 Acc.	61.0%	74.5%	79.3%	70.6%	67.6%

investigation from software to hardware, from circuit level to system level, is carried out for a complete analysis of FPGA-based deep learning accelerators and serves as a guide to future work.

3.1   WHY CHOOSE FPGAS FOR DEEP LEARNING?

Recent research works on neural networks demonstrate great improvements over traditional algorithms in machine learning. Various network models, like CNNs, recurrent neural networks (RNNs), have been proposed for image, video, and speech processes. CNNs [74] improve the top-5 image classification accuracy on ImageNet [75] dataset from 73.8–84.7% in 2012 and further improve object detection [76] with its outstanding ability in feature extraction. RNNs [77] achieve the state-of-the-art word error rate on speech recognition. In general, NNs feature a high fitting ability to a wide range of pattern recognition problems. This ability makes NNs promising candidates for many artificial intelligence applications.

But the computation and storage complexity of NN models are high. In Table 3.1, we list the number of operations, number of parameters (add or multiplication), and top-1 accuracy on ImageNet dataset [75] of state-of-the-art CNN models. Take CNNs as an example. The largest CNN model for a 224×224 image classification requires up to 39 billion floating-point operations (FLOP) and more than 500 MB model parameters [78]. As the computational complexity is proportional to the input image size, processing images with higher resolutions may need more than 100 billion operations. Latest works like MobileNet [79] and ShuffleNet [80] are trying to reduce the network size with advanced network structures, but with obvious accuracy loss. The balance between the size of NN models and accuracy is still an open question today. In some cases, the large model size hinders the application of NNs, especially in power-limited or latency-critical scenarios.

Therefore, choosing a proper computation platform for neural-network-based applications is essential. A typical CPU can perform 10–100 GFLOP per second, and the power efficiency is usually below 1 GOP/J. So CPUs are hard to meet the high-performance requirements in cloud applications nor the low power requirements in mobile applications. In contrast, GPUs offer up to 10 TOP/s peak performance and are good choices for high-performance neural network applications. Development frameworks like Caffe [81] and Tensorflow [82] also offer easy-to-use interfaces, which makes GPUs the first choice of neural network acceleration.

Besides CPUs and GPUs, FPGAs are becoming a platform candidate to achieve energy-efficient neural network processing. With a neural network-oriented hardware design, FPGAs can implement high parallelism and make use of the properties of neural network computation to remove additional logic. Algorithm researches also show that an NN model can be simplified in a hardware-friendly way while not hurting the model accuracy. Therefore, FPGAs are capable of achieving higher energy efficiency compared with CPUs and GPUs.

FPGA-based accelerator designs are faced with two challenges in performance and flexibility.

- Current FPGAs usually support working frequency at 100–300 MHz, which is much less than CPUs and GPUs. FPGAs' logic overhead for reconfigurability also reduces the overall system performance. A straightforward design on FPGA is hard to achieve high performance and high energy efficiency.

- Implementations of neural networks on FPGAs are much harder than those on CPUs or GPUs. Development frameworks like Caffe and Tensorflow for CPUs and GPUs are currently absent for FPGAs.

Many designs addressing the above two problems have been carried out to implement energy-efficient and flexible FPGA-based neural network accelerators. In this chapter, we summarize the techniques proposed in these work from the following aspects.

- We first give a simple FPGA-based neural network accelerator performance model to analyze the methodology in energy-efficient design.

- We investigate current technologies for high-performance and energy-efficient neural network accelerator designs. We introduce the techniques at both the software and hardware level and estimate the effect of these techniques.

- We compare state-of-the-art neural network accelerator designs to evaluate the techniques introduced and estimate the achievable performance of FPGA-based accelerator design, which is at least 10× better energy efficiency than current GPUs.

- We investigate state-of-the-art automatic design methods of FPGA-based neural network accelerators.

3.2   PRELIMINARY: DEEP NEURAL NETWORK

In this section, we introduce the basic functions in neural networks. Here we only focus on the inference of NNs, which means using a trained model to predict or classify new data. The

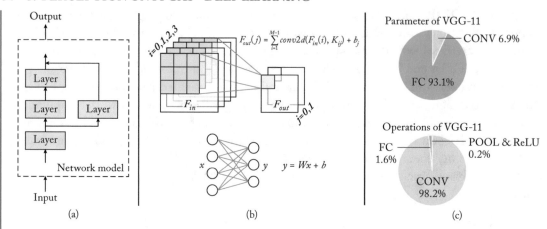

Figure 3.1: (a) Computation graph of a neural network model. (b) CONV and FC layers in NN model. (c) CONV and FC layers dominate the computation and parameter of a typical NN model: VGG11.

training process of NNs is not discussed. A neural network model can be expressed as a directed graph, as shown in Fig. 3.1a. Each vertex of the graph denotes a layer that conducts operations on data from a previous layer or input and generates results to the next layer or output. We refer to the parameter of each layer as weights and the input/output of each layer as activations through this chapter.

Convolution (CONV) layers and fully connected (FC) layers are two common types of layers in NN models. The functions of these two layers are shown in Fig. 3.1b. CONV layers conduct 2D convolutions on a set of input feature maps F_{in} and add the results to get output feature maps F_{out}. FC layers receive a feature vector as input and conduct matrix-vector multiplications.

Besides CONV and FC layers, NN layers also have pooling, ReLU [74], concat [83], element-wise [84], and other types of layers. But these layers contribute little to the computation and storage requirement of a neural network model. Figure 3.1c shows the distribution of weights and operations in the VGG-11 model [78]. In this model, CONV and FC layers together contribute more than 99% of the network's weights and operations, which is similar to most of the CNN models. Compared with CNN, RNN models [77, 85] usually have no CONV layers, and only FC layers contribute to most of the computation and storage. So most of the neural network acceleration systems focus on these two types of layers.

3.3   DESIGN METHODOLOGY AND CRITERIA

Before going into the details of the techniques used for neural network accelerators, we first give an overview of the design methodology. In general, the design target of a neural network

Table 3.2: List of symbols

Symbol	Description	Unit
IPS	Throughput of the system, measured by the number of inference processed each second	s^{-1}
W	Workload for each inference, measured by the number of operations* in the network, mainly addition and multiplication for neural network	GOP
OPS_{peak}	Peak performance of the accelerator, measured by the maximum number of operations can be processed each second	GOP/s
OPS_{act}	Run-time performance of the accelerator, measured by the number of operations processed each second	GOP/s
η	Utilization ratio of the computation units, measured by the average ratio of working computation units in all the computation units during each inference	—
f	Working frequency of the computation units	GHz
P	Number of computation units in the hardware design	—
L	Latency for processing each inference	s
C	Concurrency of the accelerator, measured by the number of inference processed in parallel	—
Eff	Energy efficiency of the system, measured by the number of operations can be processed within unit energy	GOP/J
E_{total}	Total system energy cost for each inference	J
E_{static}	Static energy cost of the system for each inference	J
E_{op}	Average energy cost for each operation in each inference	J
N_{x_acc}	Number of bytes accessed from memory (x can be SRAM or DRAM)	byte
E_{x_acc}	Energy for accessing each byte from memory (x can be SRAM or DRAM)	J/byte

*Each addition or multiplication is counted as one operation.

inference accelerator includes the following two aspects: high speed (high throughput and low latency) and high energy efficiency. The symbols used in this section are listed in Table 3.2.

Speed. The throughput of an NN accelerator can be expressed by Eq. (3.1). The on-chip resource for a certain FPGA chip is limited. We can increase the peak performance by: (1) increasing the number of computation units P by reducing the size of each computation unit and (2) increasing the working frequency f. Reducing the size of computation units can be achieved by sacrificing the data precision, which may hurt the model accuracy and requires hardware-software co-design. On the other hand, increasing working frequency is pure hardware design work. Corresponding techniques on software models and hardware are introduced in Sections 3.4 and 3.5, respectively. A high utilization ratio η is ensured by reasonable parallelism implementation and an efficient memory system. The property of the target model, i.e., the data access pattern or data-computation ratio, also affects if the hardware can be fully utilized at run-time. But most of the previous work targeting a higher utilization ratio focus on the hardware side.

$$IPS = \frac{OPS_{act}}{W} = \frac{OPS_{peak} \times \eta}{W} = \frac{fP \times \eta}{W}. \tag{3.1}$$

Most of the FPGA-based NN accelerators compute different inputs one by one. Some designs process different inputs in parallel. So the latency of the accelerator is expressed as Eq. (3.2). The common concurrent design includes layer pipeline and batch processing. This is usually considered together with loop unrolling and will be introduced in Section 3.5.2. In this chapter, we focus more on optimizing the throughput. As different accelerators may be evaluated on different NN models, a common criterion of speed is the OPS_{act}, which eliminates the effect of different network models to some extent.

$$L = \frac{C}{IPS}. \tag{3.2}$$

Energy Efficiency. Energy efficiency (Eff) is another critical criterion for computing systems. For neural network inference accelerators, energy efficiency is defined as Eq. (3.3). Like throughput, we count the number of operations rather than the number of inference to eliminates the difference of workload W. If the workload for the target network is fixed, increasing the energy efficiency of a neural network accelerator means reducing the total energy cost, E_{total} to process each input.

$$Eff = \frac{W}{E_{total}} \tag{3.3}$$

$$E_{total} \approx W \times E_{op} + N_{SRAM\_acc} \times E_{SRAM\_acc}$$
$$+ N_{DRAM\_acc} \times E_{DRAM\_acc} + E_{static}. \tag{3.4}$$

The total energy cost mainly comes from two parts: computation and memory access, which are expressed in Eq. (3.4). The first item in Eq. (3.4) is the dynamic energy cost for computation. Given a certain network, the workload W is fixed. Researchers have been focusing on optimizing the NN models by quantization (narrowing the bit-width used for computation) to reduce E_{op} or sparsification (setting more weights to zeros) to skip the multiplications with these zeros to reduce N_{op}, which follows similar rules as for throughput optimization.

The second and third items in Eq. (3.4) are the dynamic energy cost for memory access. An FPGA-based NN accelerator usually works with an external DRAM. We separate the memory access energy into the DRAM part and SRAM part. N_{x_acc} can be reduced by quantization, sparsification, efficient on-chip memory system, and scheduling method. Thus, these methods help reduce dynamic memory energy. Corresponding methods will be introduced in Section 3.5.3. The unit energy E_{x_acc} can hardly be reduced given a certain FPGA platform.

The fourth item E_{static} denotes the static energy cost of the system. This energy cost can hardly be improved given the FPGA chip and the scale of the design.

From the analysis of speed and energy, we see that the neural network accelerator involves both optimizations on NN models and hardware. In the following sections, we will introduce previous work in these two aspects, respectively.

3.4 HARDWARE-ORIENTED MODEL COMPRESSION

As introduced in Section 3.3, the design of energy-efficient and fast neural network accelerators can benefit from the optimization of NN models. A larger NN model usually results in higher model accuracy. This means it is possible to trade the model accuracy for the hardware speed or energy cost. Neural network researchers are designing more efficient network models from AlexNet [74] to ResNet [84], SqueezeNet [86], and MobileNet [79]. The latest work tries to directly optimize the processing latency by searching a good network structure [87] or skip some layers at run-time to save computation [88]. Also, there are recent approaches to achieve high performance and low memory footprint through compression-compilation co-designs [89]. Within these methods, the main differences between the handcrafted/generated networks are the size and the connections between each layer. The basic operations are the same and the differences hardly affect the hardware design. For this reason, we will not focus on these techniques in this chapter. But designers should consider using these techniques to optimize the target network.

Other methods try to achieve the tradeoff by compressing existing NN models. They try to reduce the number of weights or reduce the number of bits used for each activation or weight, which helps lower down the computation and storage complexity. Corresponding hardware designs can benefit from these NN model compression methods. In this section, we investigate these hardware-oriented network model compression methods.

3.4.1 DATA QUANTIZATION

One of the most commonly used methods for model compression is the quantization of the weights and activations. The activations and weights of a neural network are usually represented by floating-point data in common developing frameworks. Recent work tries to replace this representation with low-bit fixed-point data or even a small set of trained values. On the one hand, using fewer bits for each activation or weight helps reduce the bandwidth and storage requirement of the neural network processing system. On the other hand, using a simplified representation reduce the hardware cost for each operation. The benefit of hardware will be discussed in detail in Section 3.5. Two kinds of quantization methods are discussed in this section: linear quantization and nonlinear quantization.

Linear Quantization

Linear quantization finds the nearest fixed-point representation of each weight and activation. The problem with this method is that the dynamic range of floating-point data greatly exceeds that of fixed-point data. Most of the weights and activations will suffer from overflow or underflow. Qiu et al. [90] find that the dynamic range of the weights and activations in a single layer is much more limited and differs across different layers. Therefore, they assign different fractional bit-widths to the weights and activations in different layers. To decide the fractional bit-width of a set of data, i.e., the activations or weights of a layer, the data distribution is first analyzed. A set of possible fractional bit-widths are chosen as candidate solutions. Then the solution with the best model performance on the training data set is chosen. In [90], the optimized solution of a network is chosen layer by layer to avoid an exponential design space exploration. Wang et al. [91] try to use large bit-width for only the first and last layer and quantize the middle layers to ternary or binary. The method needs to increase the network size to keep high accuracy but still brings hardware performance improvement. Guo et al. [92] choose to fine-tune the model after the fractional bit-width of all the layers are fixed. Tambe et al. [93, 94] present an algorithm-hardware co-design centered around a novel floating-point inspired number format that dynamically maximizes and optimally clips its available dynamic range, at a layer granularity, in order to create faithful encodings of neural network parameters. Krishnan et al. [95] apply post-training quantization and quantization-aware training techniques on reinforcement learning tasks.

The method of choosing a fractional bit-width equals to scale the data with a scaling factor of 2^k. Li et al. [96] scale the weights with trained parameter W^l for each layer and quantize the weights with 2-bit data, representing W^l, 0, and $-W^l$. The activations in this work are not quantized. So the network still implements 32-bit floating-point operations. Zhou et al. [97] further quantize the weights of a layer with only 1 bit to $\pm s$, where $s = E(|w^l|)$ is the expectation of the absolute value of the weights of this layer. Linear quantization is also applied to the activations in this work.

Nonlinear Quantization

Compared with linear quantization, nonlinear quantization independently assigns values to different binary codes. The translation from a nonlinear quantized code to its corresponding value is thus a look-up table. This kind of method helps further reduce the bit-width used for each activation or weight. Chen et al. [98] assign each of the weights to an item in the look-up table by a pre-defined hash function and train the values in look-up tables. Han et al. [99] assign the values in look-up tables to the weights by clustering the weights of a trained model. Each look-up table value is set as the cluster center and further fine-tuned with the training data set. This method can compress the weights of state-of-the-art CNN models to 4-bit without accuracy loss. Zhu et al. [100] propose the ternary-quantized network where all the weights of a layer are quantized to three values: W^n, 0, and W^p. Both the quantized value and the correspondence

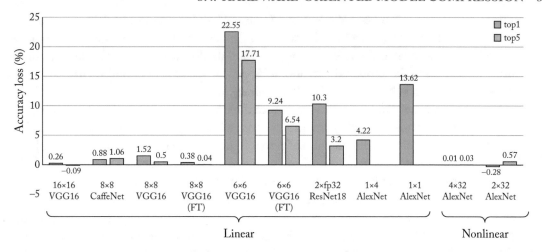

Figure 3.2: Comparison between different quantization methods from [90, 92, 96, 97, 99, 100]. The quantization configuration is expressed as (weight bit-width) × (activation bit-width). The "(FT)" denotes that the network is fine-tuned after a linear quantization.

between weights and the look-up table are trained. This method sacrifices less than 2% accuracy loss on ImageNet dataset on state-of-the-art network models. The weight bit-width is reduced from 32-bit to 2-bit, which means about 16× model size compression.

Comparison

We compare some typical quantization methods from [90, 92, 96, 97, 99, 100] in Fig. 3.2. All the quantization results are tested on ImageNet data set, and the absolute accuracy loss compared with corresponding baseline floating-point models is recorded. Comparing different methods on different models is a little bit unfair. But it still gives some insights. For linear quantization, 8-bit is a clear bound to ensure negligible accuracy loss. With 6 or fewer bits, using fine-tune or even training each weight from the beginning will cause noticeable accuracy degradation. If we require that 1% accuracy loss is within the acceptable range, linear quantization with at least 8 × 8 configuration and the listed nonlinear quantization are available. We will further discuss the performance gain of quantization in Section 3.5.

3.4.2   WEIGHT REDUCTION

Besides narrowing the bit-width of activations and weights, another method for model compression is to reduce the number of weights. One kind of method is to approximate the weight matrix with a low-rank representation. Qiu et al. [90] compress the weight matrix W of an FC layer with singular value decomposition. An $m \times n$ weight matrix W is replaced by the multiplication of two matrices $A_{m \times p} B_{p \times n}$. For a sufficiently small p, the total number of weights is reduced. This work compresses the largest FC layer of VGG network to 36% of its original size with 0.04% classification accuracy degradation. Zhang et al. [101] use a similar method for convolution layers and takes the effect of the following nonlinear layer into the decomposition optimization process. The proposed method achieves 4× speed up on state-of-the-art CNN model targeting at ImageNet, with only 0.9% accuracy loss.

Pruning is another kind of method to reduce the number of weights. This kind of method directly removes the zeros in weights or removes those with small absolute values. The challenge in pruning is the tradeoff between the ratio of zero weights and the model accuracy. One solution is the application of the lasso method, which applies L1 normalization to the weights during training. Liu et al. [102] apply the sparse group-lasso method on the AlexNet [74] model. 90% weights are removed after training with less than 1% accuracy loss. Another solution is to prune the zero weights during training. Han et al. [99] directly remove the weights of a network that are zero or have small absolute value. The left weights are then fine-tuned with the training dataset to recover accuracy. Experimental results on AlexNet show that 89% weights can be removed while keeping the model accuracy.

The hardware gain from weight reduction is the reciprocal of the compression ratio. According to the above results, the potential speed improvement from weight reduction is up to 10×.

3.5 HARDWARE DESIGN: EFFICIENT ARCHITECTURE

In this section, we investigate the hardware level techniques used in state-of-the-art FPGA-based neural network accelerator design to achieve high performance and high energy efficiency. We classify the techniques into three levels: computation unit level, loop unrolling level, and system level.

3.5.1 COMPUTATION UNIT DESIGNS

Computation unit-level design affects the peak performance of the neural network accelerator. The available resource of an FPGA chip is limited. A smaller computation unit design means more computation units and higher peak performance. A carefully designed computation unit array can also increase the working frequency of the system and thus improve peak performance.

Low Bit-Width Computation Unit
Reducing the number of bit-width for computation is a direct way to reduce the size of computation units. The feasibility of using fewer bits comes from the quantization methods as introduced in Section 3.4.1. Most of the state-of-the-art FPGA designs replace the 32-bit floating-point units with fixed-point units. Podili et al. [103] implement 32-bit fixed-point units for the proposed system. 16-bit fixed-point units are widely adopted in [90, 104–107]. ESE [108] adopts 12-bit fixed-point weight and 16-bit fixed-point neurons design. Guo et al. [92] use 8-bit units for their design on embedded FPGA. Recent work also focuses on extremely narrow bit-width

Table 3.3: FPGA resource consumption comparison for multiplier and adder with different types of data

| | Xilinx Logic | | | | Xilinx DSP | | | Altera DSP | |
| | Multiplier | | Adder | | Multiply and Add | | | Multiply and Add | |
	LUT	FF	LUT	FF	LUT	FF	DSP	ALM	DSP
fp32	708	858	430	749	800	1284	2	1	1
fp16	221	303	211	337	451	686	1	213	1
fixed32	1112	1143	32	32	111	64	4	64	3
fixed16	289	301	16	16	0	0	1	0	1
fixed8	75	80	8	8	0	0	1	0	1
fixed4	17	20	4	4	0	0	1	0	1

design. Prost-Boucle et al. [109] implements 2-bit multiplication with 1 LUT for ternary networks. Experiments in [110] show that FPGA implementation of Binarized Neural Network (BNN) outperforms that on CPU and GPU. Though BNN suffers from accuracy loss, many designs explore the benefit of using 1-bit data for computation [111–119].

The designs mentioned above focus on computation units for linear quantization. For nonlinear quantization, translating the data back to full precision for computation still costs many resources. Samragh et al. [120] propose the factorized coefficients based dot product implementation. As the possible values of weights are quite limited for nonlinear quantization, the proposed computation unit accumulates the multipliers for each possible weight value and calculates the result as the weighted sum of the values in look-up tables. In this way, the multiplication needed for one output neuron equals the number of values in the look-up table. The multiplications are replaced by random-addressed accumulations.

Most of the designs use one bit-width through the process of a neural network. Qiu et al. [90] find that neurons and weights in FC layers can use fewer bits compared with CONV layers while the accuracy is maintained. Heterogeneous computation units are used in the designs of [113, 121].

The size of computation units of different bit-widths is compared in Table 3.3. Three kinds of implementations are tested: separate multiplier and adder with the logic resource on Xilinx FPGA, multiply-add function with DSP units on Xilinx FPGA, and multiply-add function with DSP units on Altera FPGA. The resource consumption is the synthesis result by Vivado 2018.1 targeting Xilinx XCKU060 FPGA and Quartus Prime 16.0 targeting Altera Arria 10 GX1150 FPGA. The pure logic modules and the floating-point multiply and add modules are generated with IP core. The fixed-point multiply and add modules are implemented with $A * B + C$ in Verilog and automatically mapped to DSP by Vivado/Quartus.

We first give an overview of the size of the computation units by logic-only implementations. By compressing the weights and activations from 32-bit floating-point number to 8-bit fixed-point number, the multiplier and the adder are scaled down to about 1/10 and 1/50, respectively. Using 4-bit or smaller operators can bring further advantage but also incur significant accuracy loss as introduced in Section 3.4.1.

Recent FPGAs consist of a large number of DSP units, each of which implements a hard multiplier, pre-adder, and accumulator core. The basic pattern of NN computation, multiplication and sum, also fits into this design. So we also test the multiply and add function implemented with DSP units. Because of the different DSP architectures, we test on both Xilinx and Altera platforms. Compared with the 32-bit floating-point function, fixed-point functions with narrow bit-width still shows an advantage in resource consumption. But for Altera FPGA, this advantage is not obvious because the DSP units natively support floating-point operations.

Fixed-point functions with 16-or-less-bit fixed-point data are well fit into 1 DSP unit on either Xilinx or Altera FPGA. This shows that quantization hardly benefits the hardware if we use narrower bit-width like 8 or 4 in the aspect of computation. The problem is that the wide multipliers and adders in DSP units are underutilized in these cases. Nguyen et al. [122] propose the design to implement two narrow bit-width fixed-point multiplication with a single-wide bit-width fixed-point multiplier. In this design, two multiplications, AB and AC, are executed in the form of $A(B << k + C)$. If k is sufficiently large, the bits for AB and AC do not overlap in the multiplication result and can be directly separated. The design in [122] implements two 8-bit multiplications with one 25×18 multiplier, where k is 9. Similar methods can be applied to other bit-width and DSPs.

Fast Convolution Method

For CONV layers, the convolution operations can be accelerated by alternative algorithms. Discrete Fourier Transformation- (DFT) based fast convolution is widely adopted in digital signal processing. Zhang et al. [123] propose a 2D DFT-based hardware design for efficient CONV layer execution. For an $F \times F$ filter convolved with $K \times K$ filter, DFT converts the $(F - K + 1)^2 K^2$ multiplications in the space domain to F^2 complex multiplications in the frequency domain. For a CONV layer with M input channel and N output channel, MN times of frequency domain multiplications and $(M + N)$ times DFT/IDFT are needed. The conversion of convolution kernels is once for all. So the domain conversion process is of low cost for CONV layers. This technique does not work for CONV layers with stride > 1 or 1×1 convolution. Ding et al. [124] suggest that a block-wise circular constraint can be applied to the weight matrix. In this way, the matrix-vector multiplication in FC layers is converted to a set of 1D convolutions and can be accelerated in the frequency domain. This method can also be applied to CONV layers by treating the $K \times K$ convolution kernels as $K \times K$ matrices and is not limited by K or stride.

Frequency domain methods require complex number multiplication. Another kind of fast convolution involves only real number multiplication [125]. The convolution of a 2D feature map F_{in} with a kernel K using Winograd algorithm is expressed by Eq. (3.5):

$$F_{out} = A^T[(GF_{in}G^T) \odot (BF_{in}B^T)]A \tag{3.5}$$

G, B, and A are a transformation matrix which only related to the sizes of kernel and feature map. $\odot$ denotes an element-wise multiplication of two matrices. For a 4×4 feature map convolved with a 3×3 kernel, the transformation matrices are described as follows:

$$G = \begin{bmatrix} 1 & 0 & 0 \\ \frac{1}{2} & \frac{1}{2} & \frac{1}{2} \\ \frac{1}{2} & -\frac{1}{2} & \frac{1}{2} \\ 0 & 0 & 1 \end{bmatrix} \quad B = \begin{bmatrix} 1 & 0 & -1 & 0 \\ 0 & 1 & 1 & 0 \\ 0 & -1 & 1 & 0 \\ 0 & 1 & 0 & -1 \end{bmatrix} \quad A = \begin{bmatrix} 1 & 0 \\ 1 & 1 \\ 1 & -1 \\ 0 & -1 \end{bmatrix}$$

Multiplication with transformation matrices A, B, and G induce only a small number of shifts and additions because of the special matrix entries. In this case, the number of multiplication is reduced from 36 to 16. The most commonly used Winograd transformation is for 3×3 convolutions in [105, 126].

The theoretical performance gain from fast convolution depends on the convolution size. Limited by the on-chip resource and the consideration of flexibility, current designs are not choosing large convolution sizes. Existing works point out that up to 4× theoretical performance gain can be achieved by fast convolution with FFT [123] or Winograd [126] with reasonable kernel sizes. Zhuge et al. [127] even try to use both FFT and Winograd methods in their design to fit different kernel sizes in different layers.

Frequency Optimization Methods

All the above techniques introduced targets at increasing the number of computation units within a certain FPGA. Increasing the working frequency of the computation units also improves the peak performance.

The latest FPGAs support 700–900 MHz DSP theoretical peak working frequency. But existing designs usually work at 100–400 MHz [90, 92, 107, 128, 129]. As claimed in [130], the working frequency is limited by the routing between on-chip SRAM and DSP units. The design in [130] uses different working frequencies for DSP units and surrounding logic. Neighbor slices to each DSP unit are used as local RAMs to separate the clock domain. The prototype design in [130] achieves the peak DSP working frequency at 741 MHz and 891 MHz on FPGA chips of different speed grades. Xilinx has also proposed the CHaiDNN-v2 [131] and xfDNN [132] with this technique and achieves up to 700 MHz DSP working frequency. Compared with existing designs for which the frequency is within 300 MHz, this technique brings at least 2× the peak performance gain.

3.5.2   LOOP UNROLLING STRATEGIES

CONV layers and FC layers contribute to most of the computations and storage requirements of a neural network. We express the CONV layer function in Fig. 3.1b as nested loops in Algorithm 3.1. To make the code clear to read, we merge the loops along x and y directions for feature maps and 2D convolution kernels, respectively. An FC layer can be expressed as a CONV layer with feature map and kernel both of size 1×1. Besides the loops in Algorithm 3.1, we also call the parallelism of the process of multiple inputs as a batch. As we treat FC layers and CONV layers all as nested loops, the loop unrolling strategy can be applied both in CNN accelerators and RNN accelerators. But as the case for FC layers is rather simple, we tend to use CNN as examples in this section.

Algorithm 3.1 Convolution Layer

Require: feature map F_{in} of size $M \times Y \times X$; convolution kernel Ker of size $N \times M \times K \times K$; bias vector b of size N

Ensure: feature map F_{out}

Function ConvLayer(F_{in}, Ker)

1: Let $F_{out} \leftarrow$ zero array of size $N \times (Y - K + 1) \times (X - K + 1)$
2: **for** $n = 1; n < N; n + +$ **do**
3:     **for** $m = 1; m < M; m + +$ **do**
4:        **for** each (y, x) within $(Y - K + 1, X - K + 1)$ **do**
5:           **for** each (ky, kx) within (K, K) **do**
6:              $F_{out}[n][y][x] + = F_{in}[m][y - ky + 1][x - kx + 1] * K[n][m][ky][kx]$
7:           **end for**
8:        **end for**
9:     **end for**
10:    $F_{out}[n] + = b[n]$
11: **end for**
12: **return** F_{out}

Choosing Unroll Parameters

To parallelize the execution of the loops, we unroll the loops and parallelize the process of a certain number of iterations on hardware. The number of the parallelized iterations on hardware is called the unroll parameter. Inappropriate unroll parameter selection may lead to serious hardware underutilization. Take a single loop as an example. Suppose the trip count of the loop is M and the parallelism is m. The utilization ratio of the hardware is limited by $m / M \lceil M/m \rceil$. If M is not divisible by m, then the utilization ratio is less than 1. For processing an NN layer, the total utilization ratio will be the product of the utilization ratio on each of the loops.

For a CNN model, the loop dimension varies greatly among different layers. For a typical network used on ImageNet classification like ResNet [84], the channel numbers vary from 3 to 2048, the feature map sizes vary from 224×224 to 7×7, and the convolution kernel sizes vary from 7×7 to 1×1. Besides the underutilization problem, loop unrolling also affects the datapath and on-chip memory design. Thus, the loop unrolling strategy is a key feature for a neural network accelerator design.

Various works are proposed focusing on how to choose the unroll parameters. Zhang et al. [133] propose the idea of unrolling the input channel and output channel loops and choose the optimized unroll parameter by design space exploration. Along with these two loops, there is no input data cross-dependency between neighboring iterations. So, no multiplexer is needed to route data from the on-chip buffer to computation units. But the parallelism is limited as $7 \times 64 = 448$ multipliers. For larger parallelism, this solution is easy to suffer from the underutilization problem. Ma et al. [128] further extend the design space by allowing parallelism on the feature map loop. The parallelism reaches $1 \times 16 \times 14 \times 14 = 3136$ multipliers. A shift register structure is used to route feature map pixels to the computation units.

The kernel loop is not chosen in the above work because kernel sizes vary greatly. Motamedi et al. [134] use kernel unrolling on AlexNet. Even with 3×3 unrolling for the 11×11 and 5×5 kernels, the overall system performance still reaches 97.4% of its peak performance for the convolution layers. For certain networks like VGG [78], only 3×3 convolution kernels are used. Another reason to unroll the kernel loop is to achieve acceleration with fast convolution algorithms. Design in [123] implements fully parallelized frequency domain multiplication on 4×4 feature map and 3×3 kernel. Lu et al. [126] implement Winograd algorithm on FPGA with a dedicated pipeline for Eq. (3.5). The convolution of a 6×6 feature map with a 3×3 kernel is fully parallelized.

The above solutions are only for a single layer. But there is hardly a one-size-fits-all solution for a whole network, especially when we need high parallelism. Designs in [104, 135, 136] propose fully pipelined structures with each layer a pipe stage. As each layer is executed with an independent part of the hardware and each part is small, the loop unrolling method can be easily chosen. This method is memory-consuming because ping-pong buffers are needed between adjacent layers for the feature maps. Aggressive design with binarized weights [118] can fit into FPGA better. Design in [137] is similar but implemented on FPGA clusters to resolve the scalability problem. Shen et al. [138] and Lin et al. [139] group the layers of a CNN by the loops' trip count and map each group onto one hardware module. These solutions can be treated as unrolling the batch loop because different inputs are processed in parallel on different layer pipeline stages. The design in [126] implements parallelized batch both within a layer and among different layers.

Most of the current designs follow one of the above methods for loop unrolling. A special kind of design is for sparse neural networks. Han et al. [108] propose the ESE architecture for sparse long short-term memory (LSTM) network acceleration. Unlike processing a dense network, all the computation units will not work synchronously. This causes difficulty in sharing data between different computation units. ESE [108] implements only the output channel (the output neurons of the FC layers in LSTM) loop unrolling within a layer to simplify hardware design and parallelize the batch process.

Data Transfer and On-chip Memory Design

Besides the high parallelism, the on-chip memory system should efficiently offer the necessary data to each computation units every cycle. To achieve high parallelism, neural network accelerators usually reuse data among a large number of computation units. Simply broadcasting data to different computation units leads to large fan-out and high routing costs, thus reducing the working frequency. Wei et al. [140] use the systolic array structure in their design. The shared data are transferred from one computation unit to the next in a chain mode. So the data is not broadcasted, and only local connections between different computation units are needed. The drawback is the increase in latency. The loop execution order is scheduled accordingly to cover the latency. Similar designs are adopted in [128, 141].

For software implementation on GPU, the im2col function is commonly used to map 2D convolution as a matrix-vector multiplication. This method incurs considerable data redundancy and can hardly be applied to the limited on-chip memory of FPGAs. Qiu et al. [90] uses the line buffer design to achieve the 3×3 sliding window function for 2D convolution with only two lines of duplicated pixels.

3.5.3   SYSTEM DESIGN

A typical FPGA-based neural network accelerator system is shown in Fig. 3.3. The logic part of the whole system is denoted by the blue boxes. The host CPU issues workload or commands to the FPGA logic part and monitors its working status. On the FPGA logic part, a controller is usually implemented to communicate with the host and generates control signals to all the other modules on FPGA. The controller can be an FSM or an instruction decoder. The on the fly logic part is implemented for certain designs if the data loaded from external memory needs preprocess. This module can be data arrangement module, data shifter [90], FFT module [123], etc. The computation units are discussed in Sections 3.5.1 and 3.5.2. As the on-chip SRAM of an FPGA chip is too limited compared with the large NN models, a two-level memory hierarchy is used with DDR and on-chip memory for common designs.

Roofline Model

From the system level, the performance of a neural network accelerator is limited by two factors: the on-chip computation resource and the off-chip memory bandwidth. Various researches have been proposed to achieve the best performance within a certain off-chip memory bandwidth. Zhang et al. [133] introduce the roofline model in their work to analyze whether a design is memory bounded or computation bounded. An example of a roofline model is shown in Fig. 3.4.

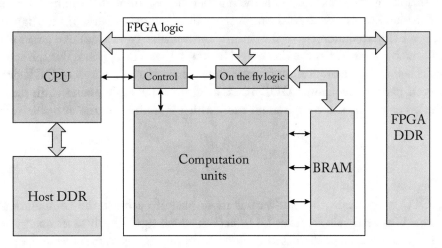

Figure 3.3: Block graph of a typical FPGA-based neural network accelerator system.

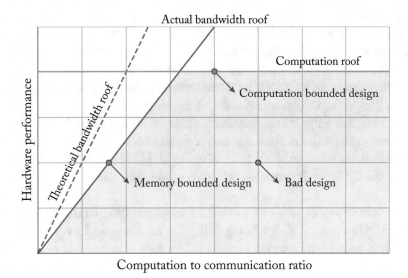

Figure 3.4: An example of the roofline model. The shaded part denotes the valid design space given bandwidth and resource limitation.

The figure uses the computation to communication (CTC) ratio as the x-axis and hardware performance as the y-axis. CTC is the number of operations that can be executed with a unit size of memory access. Each hardware design can be treated as a point in the figure. So y/x equals the bandwidth requirement of the design. The available bandwidth of a target platform is limited and can be described as the theoretical bandwidth roof in Fig. 3.4. But the actual bandwidth roof is below the theoretical roof because the achievable bandwidth of DDR depends on the data access pattern. Sequential DDR access achieves much higher bandwidth than random access. The other roof is the computation roof, which is limited by the available resource on FPGA.

Loop Tiling and Interchange

A higher CTC ratio means the hardware is more likely to achieve the computation bound. Increasing the CTC ratio also reduces DDR access, which significantly saves energy according to [142]. In Section 3.5.2, we discussed the loop unrolling strategies to increase the parallelism while reducing the waste of computation for a certain network. When the loop unrolling strategy is decided, the scheduling of the rest part of the loops decides how the hardware can reuse data with on-chip buffer. This involves loop tiling and loop interchange strategy.

Loop tiling is a higher level of loop unrolling. All the input data of a loop tile will be stored on-chip, and the loop unrolling hardware kernel works on these data. A larger loop tile size means that each tile will be loaded from external memory to on-chip memory fewer times. Loop interchange strategy decides the processing order of the loop tiles. External memory access happens when the hardware is moving from one tile to the next. Neighboring tiles may share a part of data. For example, in a CONV layer, neighboring tiles can share input feature maps or weights. This is decided by the execution order of the loops.

In [128, 133], design space exploration is done on all the possible loop tiling sizes and loop orders. Many designs also explore the design space with some of the loop unrolling, tiling, and loop order is already decided [90, 134]. Shen et al. [143] also discuss the effect of batch parallelism over the CTC for different layers. This is a loop dimension not focused on in previous work.

All of the above works give one optimized loop unrolling strategy and loop order for a whole network. Guo et al. [92] implements flexible unrolling and loop order configuration for different layers with an instruction interface. The data arrangement in on-chip buffers is controlled through instructions to fit with different feature map sizes. This means the hardware can always fully utilize the on-chip buffer to use the largest tiling size according to the on-chip buffer size. This work also proposes the "back and forth" loop execution order to avoid total on-chip data refresh when an innermost loop finishes.

Cross-Layer Scheduling

Alwani et al. [144] address the external memory access problem by fusing two neighboring layers together to avoid the intermediate result transfer between the two layers. This strategy helps reduce 95% off-chip data transfer with an extra 20% on-chip memory cost. Even software program gains 2× speedup with this scheduling strategy. Yu et al. [145] realize this idea on a single-layer accelerator design by modifying the order of execution through an instruction interface.

Regularize Data Access Pattern

Besides increasing CTC, increasing the actual bandwidth roof helps improve the achievable performance with a certain CTC ratio. This is achieved by regularizing the DDR access pattern. The common feature map formats in the external memory include *NCHW* or *CHWN*, where *N* means the batch dimension, *C* means the channel dimension, *H* and *W* means the feature map *y* and *x* dimension. Using any of these formats, a feature map tile may be cut into small data blocks stored in discontinuous addresses. Guan [106] suggest that a channel-major storage format should be used for their design. This format avoids data duplication while long DDR access burst is ensured. Qiu et al. [90] propose a feature map storage format that arranges the $H \times W$ feature map into (HW/rc) tile blocks of size $r \times c$. So the write burst size can be increased from $c/2$ to $rc/2$.

3.6   EVALUATION

In this section, we compare the performance of state-of-the-art neural network accelerator designs and try to evaluate the techniques mentioned in Sections 3.4 and 3.5. We mainly reviewed the FPGA-based designs published in the top FPGA conferences (FPGA, FCCM, FPL, FPT), EDA conferences (DAC, ASP-DAC, DATE, ICCAD), architecture conferences (MICRO, HPCA, ISCA, ASPLOS) from 2015 to 2019. Because of the diversity in the adopted techniques, target FPGA chips, and experiments, we need a trade-off between the fairness of comparison and the number of designs we can use. In this chapter, we pick the designs with (1) whole system implementation and (2) experiments on real NN models with reported speed, power, and energy efficiency.

The designs used for comparison are listed in Table 3.4. For data format, the "INT A/B" means that activations are A-bit fixed-point data and weights are B-bit fixed-point data. We also investigate the resource utilization and draw advice to both accelerator designers and FPGA manufacturers. Besides the FPGA-based designs, we also plot the GPU results used in [92, 108] as standards to measure the FPGA designs' performance.

Bit-Width Reduction

Among all the designs, 1–2 bit-based designs [115–117] show outstanding speed and energy efficiency. This shows that extremely low bit-width is a promising solution for high-performance

Table 3.4: Performance and resource utilization of state-of-the-art neural network accelerator designs

ID	Ref.	Data Format	Speed (GOP/s)	Power (W)	Eff. (GOP/J)	Resource (%)			FPGA chip
						DSP	Logic	BRAM	
1	[115]	1 bit	329.47	2.3	143.2	1	34	11	XC7Z020
2	[117]	1 bit	40770	48	849.38	–	–	–	GX1155
3	[116]	2 bit	410.22	2.26	181.51	41	83	38	XC7Z020
4	[92]	INT8	84.3	3.5	24.1	87	84	89	XC7Z020
5	[146]	INT16/8	117.8	19.1	6.2	13	22	65	5SGSD8
6	[135]	INT16/8	222.1	24.8	8.96	40	27	40	XC7VX690T
7	[128]	INT16/8	645.25	21.2	30.43	100	38	70	GX1150
8	[108]	INT16/12	2520	41	61.5	54	89	88	XCKU060
9	[147]	INT16	12.73	1.75	7.27	95	67	6	XC7Z020
10	[90]	INT16	136.97	9.63	14.22	89	84	87	XC7Z045
11	[105]	INT16	229.5	9.4	24.42	92	71	83	XC7Z045
12	[107]	INT16	354	26	13.6	78	81	42	XC7VX690T
13	[106]	INT16	364.4	25	14.6	65	25	46	5SGSMD5
14	[104]	INT16	565.94	30.2	22.15	60	63	65	XC7VX690T
			431	25	17.1	42	56	52	XC7VX690T
15	[148]	INT16	785	26	30.2	53	8.3	30	XCVU440 XC7Z020+
16	[137]	INT16	1280.3	160	8	–	–	–	XC7VX690T×6
17	[129]	INT16	1790	37.46	47.8	91	43	53	GX1150
18	[126]	INT16	2940.7	23.6	124.6	–	–	–	ZCU102
19	[141]	FP16	1382	45	30.7	97	58	92	GX1150
20	[103]	INT32	229	8.04	28.5	100	84	18	Stratix V
21	[149]	FP32	7.26	19.63	0.37	42	65	52	XC7VX485T
22	[133]	FP32	61.62	18.61	3.3	80	61	50	XC7VX485T
23	[123]	FP32	123.5	13.18	9.37	88	85	64	Stratix V
17	[129]	FP32	866	41.73	20.75	87	–	46	GX1150

design. As introduced in Section 3.4.1, linear quantized 1–2-bit network models suffer from great accuracy loss. Further developing related accelerators will be of little use. More efforts should be put on the models. Even trading speed with accuracy can be acceptable considering the current hardware performance.

Besides the 1-2-bit designs, the rest of the designs adopts 32-bit floating-point data or linear quantization with 8 or more bits. According to the results in Section 3.4.1, within 1% accuracy loss can be achieved. So we think the comparison between these designs is fair in accuracy. INT16/8 and INT16 are commonly adopted. But the difference between these designs is not obvious. This is because of the underutilization of DSPs discussed in Section 3.5.1. The advantage of INT16 over FP32 is obvious except for [129], where the hard-core floating-point DSPs are utilized. To a certain extent, this shows the importance of fully utilizing the DSPs on-chip.

Fast Convolution Algorithm

Among all the 16-bit designs, [126] achieves the best energy efficiency and the highest speed with the help of the 6×6 Winograd fast convolution, which is 1.7× faster and 2.6× energy efficient than the 16-bit design in [129]. The design in [123] achieves 2× speedup and 3× energy efficiency compared with [133] where both designs use 32-bit floating-point data and FPGA with 28 nm technology node. Compare with the theoretical 4× performance gain introduced in Section 3.5.1, there is still 1.3–1.5× gap. Not all the layers can use the most optimized fast convolution method because of kernel size limitation.

System Level Optimization

The overall system optimization is not well addressed in most of the work. As this is also related to the HDL design quality, we can roughly evaluate the effect. Here we compare three designs [104, 107, 135] on the same XC7VX690T platform and try to evaluate the effect. All the three designs implement 16-bit fixed-point data formats except that [135] uses 8-bit for weights. No fast convolution or sparsity is utilized in any of the work. Even though, [104] achieves 2.5× the energy efficiency of [135]. It shows that a system-level optimization has a strong effect even comparable to the use of fast convolution algorithms.

We also investigate the resource utilization of the designs in Table 3.4. Three kinds of resources (DSP, BRAM, and logic) are considered. We plot the designs in Fig. 3.5 using two of the utilization ratio as x and y coordinate. We draw the diagonal line of each figure to show the designs' preference on hardware resources. The BRAM-DSP figure shows an obvious preference on DSP over BRAM. A similar preference appears on DSP over logic. This indicates that current FPGA designs are more likely computation bounded. FPGA manufacturers targeting neural network applications can adjust the resource allocation accordingly. Compared with that, the preference on logic and BRAM seems to be random. A possible explanation is that some of the

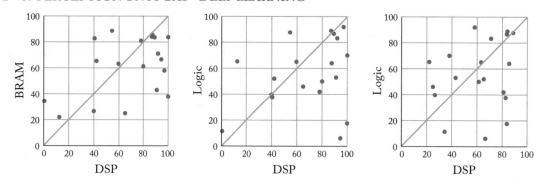

Figure 3.5: Resource utilization ratio of different accelerator designs.

designers use both logic and DSPs to implement high parallelism, while some prefer to use only DSPs to achieve high working frequency.

Comparison with GPUs

In general, FPGA-based designs have achieved comparable energy efficiency to GPU with 10–100 GOP/J. But the speed of GPUs still surpasses FPGAs. Scaling up the FPGA-based design is still a problem. Zhang et al. [137] propose the FPGA-cluster-based solution using 16-bit fixed-point computation. But the energy efficiency is worse than the other 16-bit fixed-point designs.

Here we estimate the achievable speed of an ideal design. We use the 16-bit fixed-point design in [126] as a baseline, which is the best 16-bit fixed-point design with both the highest speed and energy efficiency. 8-bit linear quantization can be adopted according to the analysis in Section 3.4.1, which achieves another 2× speedup and better energy efficiency by utilizing 1 DSP as 2 multipliers. The double frequency optimization further improves the system speed by 2×. Consider a sparse model which is similar to the one in [108] with 10% non-zero values. We can estimate a similar 6× improvement as [108]. In general, about 24× speedup and 12× better energy efficiency can be achieved, which means 72TOP/s speed with about 50 W. This shows that it is possible to achieve over 10× higher energy efficiency on FPGA over 32-bit floating-point process on GPU.

The problem left is: do all the techniques—double MAC, sparsification, quantization, fast convolution, and the double frequency design—work well together? Pruning a single element in a 2D convolution kernel is of no use for fast convolution because the 2D kernel is always processed as a whole. Directly pruning 2D kernels as a whole may help. But the reported accuracy of this method is lower [150] than a fine-grained pruning. The irregular data access pattern for processing sparse network and the increase in parallelism also brings challenges to the design of memory system and scheduling strategy.

3.7 SUMMARY

Neural networks are an essential part in robotic perception and often the main performance bottleneck. In this chapter, we have reviewed the state-of-the-art neural network accelerator designs and summarized the techniques used. We have demonstrated that with software-hardware co-design, FPGAs can achieve more than 10 times better speed and energy efficiency than the state-of-the-art GPUs. This verifies that FPGAs are a promising candidate for neural network acceleration. We have also reviewed the methods used for accelerator design automation, which shows that the current development flow can achieve both high performance and throughput. However, there are two directions for future development. On the one hand, quantization with extremely narrow bit-width is limited by the model accuracy, which needs further algorithm research. On the other hand, combining all the techniques requires more research in both software and hardware to make them work well together. Commercial tools, including DNNDK [151], are taking a first step but still has a long way to go. Scaling up the design is also a problem. We will focus on solving these challenges.

CHAPTER 4

Perception on FPGAs – Stereo Vision

Perception is related to many robotic applications where sensory data and artificial intelligence techniques are involved. The goal of perception is to sense the dynamic environment surrounding the robot and to build a reliable and detailed representation of this environment based on sensory data. Since all subsequent localization, planning, and control depends on correct perception output, its significance cannot be overstated. Perception modules usually include stereo matching, object detection, scene understanding, semantic classification, etc. The recent developments in machine learning, especially deep learning, have exposed robotic perception systems to more tasks, which have been discussed in the last chapter. In this chapter, we will focus on the algorithms and FPGA implementations in the stereo vision system, which is one of the key components in the robotic perception stage.

4.1 PERCEPTION IN ROBOTICS

This section presents an overview of the geometry-based robotic perception stage. As a core component in robotic applications, perception is understood as a system that enables the robot to perceive, comprehend and reason about the world around it. Generally, the perception stage includes object detection, object segmentation, stereo flow, and object tracking modules.

Detection

Object detection is a fundamental problem in computer vision and robotics. Fast and reliable detection of objects is crucial for performance and safety reasons. Traditionally, a detection pipeline starts with the preprocessing of input images, followed by a region of interest detector and finally a classifier that outputs detected objects. On the one hand, the object detector needs to extract distinctive features that can separate different object classes. On the other hand, it needs to construct invariant object representation that makes detection reliable.

A good object detector needs to model both the appearance and shape of the object under various conditions, and many algorithms have been proposed to address it. In 2005, Dalal and Triggs [152] proposed an algorithm based on histogram of orientation (HOG) and support vector machine (SVM). The whole algorithm is presented in Fig. 4.1. It passes input image through preprocessing, computes HOG features over sliding detection window, and uses linear

Figure 4.1: The algorithm block of HOG+SVM detection, based on Dalal and Triggs [152].

SVM classifier for detection. This algorithm captures object appearance by purposefully designed HOG feature, and depends on linear SVM to deal with highly nonlinear object articulation.

Articulated objects are challenging because of the intricate appearance of the non-rigid shape. The Deformable Part Model (DPM) introduced by Felzenszwalb et al. [153] splits objects into simpler parts so that DPM can represent non-rigid objects by composing more manageable parts. This reduces the number of training examples needed for the appearance modeling of whole objects. DPM uses HOG feature pyramid to build multiscale object hypotheses, spatial constellation model of part configuration constraint, and latent SVM to handle latent variables such as part position.

Segmentation

Segmentation can be thought of as a natural enhancement of object detection that needs to be solved sufficiently in the practical robot. Parsing images from the camera into meaningful semantic segments gives the robot a structured understanding of its environment.

Traditionally, semantic segmentation is formulated as a graph labeling problem with vertices of the graph being pixels or super-pixels. Inference algorithms on graphical models such as conditional random field (CRF) are used [154, 155]. In such an approach, CRFs are built, with vertices representing pixels or super-pixels (Fig. 4.2). Each node can take a label from a pre-defined set, conditioned on features extracted at the corresponding image position. Edges between these nodes represent constraints such as spatial smoothness, label correlations, etc.

However, CRF slows down when image dimension, input feature numbers, or label set size increases and has difficulty capturing long-range dependency in images. A highly efficient inference algorithm is proposed in [156] to improve speed for fully connected CRF with pairwise potentials between all pairs of pixels, other algorithms [157] aim to incorporate co-occurrence of object classes. Essentially, semantic segmentation needs to predict dense class labels with multi-scale image features and contextual reasoning.

Stereo and Optical Flow

Robots move in a 3D world, so the perception that produces 3D spatial information such as depth is indispensable. Similar to how humans enjoy 3D visual perception with two eyes, we can gain depth information with the stereo camera taking pictures simultaneously at slightly different angles.

Given image pair from the stereo camera (I_l, I_r), stereo is essentially a correspondence problem where pixels in left image I_l are matched to pixels in the right image I_r based on a cost

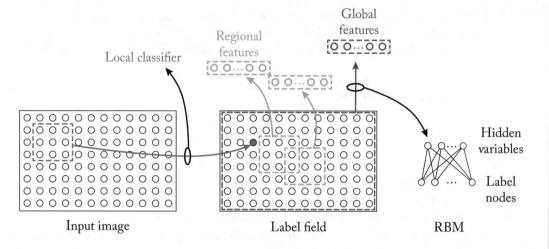

Figure 4.2: Graphical model representation of He et al. [154].

function. The assumption is that corresponding pixels map to the same physical point, thus have the same appearance:

$$I_l(p) = I_r(p + d),$$

where p is a location in the left image and d is the disparity.

Optical flow [158] is defined as 2D motion of intensities between two images, which is related but different from the 3D motion in the physical world. It relies on the same constant appearance assumption:

$$I_t(p) = I_{t+1}(p + d).$$

In stereo, image pairs are taken at the same time. Geometry is the dominating cause of the disparity, and appearance constancy is most likely to hold. In optical flow, image pairs are taken at slightly different times, and motion is just one of many varying factors such as lighting, reflections, transparency, etc.

Tracking

Tracking is responsible for estimating object states such as location, speed, and acceleration over time. Take autonomous vehicles as an example, they must track various traffic participants to maintain a safe distance and predict their trajectories.

Traditionally, tracking is formulated as a sequential Bayesian filtering problem.

1. **Prediction step:** Given object state at a previous time step, predict object state at the current time step using a motion model that describes the temporal evolution of object state.

2. **Correction step:** Given predicted object state at a current time step, and current observation from sensors, a posterior probability distribution of object state at the current time is calculated using an observation model that represents how observation is determined by object state.

3. This process goes on recursively.

4.2   STEREO VISION IN ROBOTICS

An overall picture depicting module functionality and problem formulation of general perception stage in robots application have been presented. Without losing generality, we next aim to offer a holistic view on the state-of-the-art algorithms and their FPGA-based accelerator designs on stereo vision system, a significant module of perception pipeline.

Real-time and robust stereo vision systems are increasingly popular and widely used in many applications, e.g., robotics navigation, obstacle avoidance [159], and scene reconstruction [160–162]. The purpose of stereo vision systems is to obtain 3D structure information of the scene using stereoscopic ranging techniques. The system usually has two cameras to capture images from two points of view within the same scenario. The disparities between the corresponding pixels in two stereo images are searched using stereo matching algorithms. Then the depth information can be calculated from the inverse of this disparity.

Throughout the whole stereo vision pipeline, stereo matching is the bottleneck and time-consuming stage. The stereo matching algorithms can be mainly classified into two categories: local algorithms [163–169] and global algorithms [170–174]. Local methods compute the disparities by only processing and matching the pixels around the points of interest within windows. They are fast, computationally cheap, and the lack of pixel dependencies makes them suitable for parallel acceleration. However, they may suffer in textureless areas and occluded regions, resulting in incorrect disparities estimation.

In contrast, global methods compute the disparities by matching all other pixels and minimizing a global cost function. They can achieve much higher accuracy than local methods. However, they tend to come at high computation costs and require much more resources due to their large and irregular memory access as well as the sequential nature of algorithms, thus not suitable for real-time and low-power applications.

Semi-Global Matching (SGM) [175] lies between local and global methods. It is a global method with energy function terms calculated along several 1D lines at each pixel and smoothness terms. SGM is theoretically justified [176] and also quite fast [177].

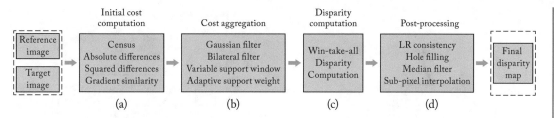

Figure 4.3: The block diagram of stereo vision system.

4.3 LOCAL STEREO MATCHING ON FPGAS

4.3.1 ALGORITHM FRAMEWORK

Generally, the local stereo vision algorithms are implemented in the following four steps: initial cost initialization, cost aggregation, disparity computation, and postprocessing. The algorithm block is presented in Fig. 4.3.

- **Step 1: Initial Cost Computation.** The dissimilarity value between a reference pixel in the left image and a set of candidate pixels in the right image over a defined disparity range is calculated (Fig. 4.3a). Many algorithms have been proposed to measure the cost, such as sum of absolute differences (SAD) [178], the sum of squared differences (SSD) [179], normalized cross-correlation (NCC) [180], and census transform (CT) [181].

- **Step 2: Cost Aggregation.** The initial costs are then aggregated in the cross-based variable support regions, which could be performed by horizontal aggregation first and vertical aggregation then. Many strategies have been proposed for cost aggregation in recent years. The most straightforward methods are low-pass filtering with fixed-sized convolution kernels such as box filters, binomial or Gaussian. Variable support window (VSW) [182] and adaptive support weight [183] (ASW)-based approaches are further proposed for various window size. The cost can be evaluated locally or globally (Fig. 4.3b).

- **Step 3: Disparity Computation.** After aggregation, the best disparity is selected with a winner-takes-all (WTA) method (Fig. 4.3c).

- **Step 4: Postprocessing.** The disparity map is further refined. The postprocessing is usually composed of outlier detection, outlier handling, and sub-pixel interpolation. Finally, a sub-pixel interpolation method is used to increase the accuracy of the disparity map (Fig. 4.3d).

Figure 4.4 shows the stereo matching windows during the disparity computation. To estimate the disparity in local stereo matching methods, the surroundings of a pixel p in the left

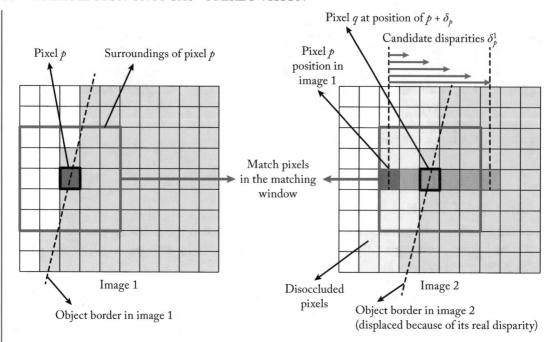

Figure 4.4: Matching windows in local stereo matching. (Figure adapted from [184].)

image are compared to the pixel q in the right image, where q has been translated over a candidate disparity δ_p compared to p. For each pixel p, N candidate disparities (δ_p^1, δ_p^2, ..., δ_p^N) are tested ($N = 256$ and 65,536 for 8- and 16-bit depth maps, respectively), and the candidate disparity resulting in the lowest matching cost is assigned to pixel p.

4.3.2   FPGA DESIGNS

Local algorithms are usually based on correlation. Some local stereo vision algorithms are suitable for software designs that make rapid algorithm updates and verification tests possible due to their high flexibility. However, they cannot exploit massive computational parallelisms and their execution time is usually higher than that of hardware counterparts. In addition, for a given local stereo vision algorithm, software design running on a general-purpose processor usually consumes much power, which is challenging to be deployed on power-constraint edge devices. Recently, many hardware designs dedicated to accelerating local stereo vision systems have been proposed and implemented.

Jin et al. [185] develop a real-time stereo vision system based on census rank transformation matching cost for 640 × 480 resolution images. Zhang et al. [186] propose a real-time high definition stereo matching design on FPGA based on mini-census transform and cross-based cost aggregation, which achieves 60 FPS at 1024×768 pixel stereo images. The implementation of Honegger et al. [187] achieves 127 FPS at 376 × 240 pixel resolution with 32 disparity levels based on block matching. Jin et al. [188] achieve 507.9 FPS for 640 × 480 resolution images by applying fast local consistent dense stereo functions and cost aggregation. Werner et al. [189] introduce an HD stereo system using NCC algorithm, which reaches 30 FPS for HD images with 60 disparity evaluations. Mattoccia et al. [190] propose a low-cost stereo system on a Xilinx Spartan-6 FPGA based on census algorithm, which achieves 30 FPS for Video Graphics Array (VGA) resolution with a disparity range of 32. Sekhar et al. [191] implement the SAD algorithm in Xilinx Zynq ZC702 SoC board and achieve 30 FPS for 640 × 360 resolution with 99 disparity levels. Perri et al. [192] propose a novel hardware-oriented stereo vision algorithm and a specific implementation suitable for heterogeneous SoC FPGA-based embedded systems. Several hardware-software partitions are characterized in this design, and a conjunct adoption of HLS- and VHDL-level descriptions. This implementation exhibits 101 FPS for 640 × 480 resolution stereo pairs.

4.4   GLOBAL STEREO MATCHING ON FPGAS

4.4.1   ALGORITHM FRAMEWORK

Compared to local algorithms, global algorithms are based on explicit smoothness assumptions. Typically, global algorithms do not perform cost aggregation step (Fig. 4.3b), but disparity computation step (Fig. 4.3c) minimizes a global cost function that combines data from initial cost computation step (Fig. 4.3a) and smoothness terms. Unlike local algorithms, global algorithms estimate the disparity at one pixel using the disparity estimates at all the other pixels.

Global algorithms can provide state-of-the-art accuracy and disparity map quality compared to local algorithms, however they are usually processed through high computational-intensive optimization techniques or massive convolutional neural networks, which makes them difficult to be deployed on resource-limited embedded systems for real-time applications. They usually rely on the high-end hardware resources of multi-core CPU or GPU platforms for execution.

4.4.2   FPGA DESIGNS

Several hardware architecture designs have been proposed to accelerate global algorithms. Yang et al. [193] propose a tree-structure global algorithm with a relatively low computational complexity. The tree filter was implemented on the CPU and achieved 20 FPS for 384 × 288 resolution image with a disparity level of 16. This algorithm could be significantly accelerated using the parallel architecture of FPGA due to its low computational complexity.

Park et al. [194] present a trellis-based stereo matching system on FPGA with a low error rate and achieved 30 FPS at 320 × 240 resolution with 128 disparity levels. Sabihuddin et al. [195] implement a dynamic programming maximum likelihood- (DPML) based hardware architecture for dense binocular disparity estimation and achieved 63.54 FPS at 640 × 480 pixel resolution with 128 disparity levels. The design in Jin et al. [196] uses a tree-structured dynamic programming method, and achieves 58.7 FPS at 640 × 480 resolution as well as a low error rate. Zha et al. [172] employ the block-based cross tree global algorithm on an FPGA platform, which is fully pipelined and parallelizes all the modules. This design can generate high accuracy disparity images of 1920 × 1680 resolution at 30 FPS with 60 disparity levels. Puglia et al. [173] implement a dynamic programming algorithm for DNA sequence alignment on FPGA. This design reached the processing ability of 1024 × 768 pixels, 64 disparity levels, and 30 FPS with a power usage on-chip of only 0.17 W, which can be used in critical and battery dependent applications. Kamasaka et al. [197] propose a parallel processing-friendly graph cut algorithm and its FPGA implementation for accelerating global stereo vision, in which object surfaces are estimated by solving a min-cut problem of a 3D grid graph. This design achieves 166× speedup compared to CPU execution of a common software library of a graph cut for 12 × 12 × 7-node graphs.

4.5   SEMI-GLOBAL MATCHING ON FPGAS

4.5.1   ALGORITHM FRAMEWORK

Semi-global matching (SGM) bridges the gap between local and global methods, and achieves a notable improvement in accuracy. SGM calculates the initial matching disparities by comparing local pixels, and then approximates an image-wide smoothness constraint with global optimization, which can obtain more robust disparity maps through this combination. There are several critical challenges for implementing SGM on hardware, e.g., data dependence, high complexity, and large storage, so this is an active research field with recent works proposing FPGA-friendly variants of SGM [177, 190, 198–200].

4.5.2   FPGA DESIGNS

Current implementations of SGM algorithms using general-purpose processors are not suitable for real-time edge applications due to their high power consumption. Therefore, FPGA-based hardware solutions are promising because of their excellent performance and low power consumption. However, there are two main challenges in implementing SGM methods on FPGAs. On the one hand, the data dependency in the optimization step of the SGM method involves the storage of a significant amount of intermediate data. On the other hand, the data access and process manner is irregular, which will impede efficient execution in hardware.

Taking these implementation challenges into consideration, several FPGA-based hardware solutions have been proposed for accelerating SGM methods. Banz et al. [200] propose a systolic-array based hardware architecture for SGM disparity estimation along with a two-dimensional parallelization concept for SGM. This design achieves 30 FPS performance at 640 × 480 pixel images with a 128-disparity range on the Xilinx Virtex-5 FPGA platform. Wang et al. [198] implement a complete real-time FPGA-based hardware system that supports absolute difference-census cost initialization, cross-based cost aggregation, and semi-global optimization. Figure 4.5 presents the overall structure of the proposed stereo matching module and detailed structure of the semi-global optimization module. The system achieves 67 FPS at 1024×768 resolution with 96 disparity levels on the Altera Stratix-IV FPGA platform, and 42 FPS at 1600×1200 resolution with 128 disparity levels on the Altera Stratix-V FPGA platform.

The design in Cambuim et al. [201] uses a scalable systolic-array based architecture for SGM based on the Cyclone IV FPGA platform, and it achieves a 127 FPS image delivering rate in 1024×768 pixel HD resolution with 128 disparity levels. The key point of this design is the combination of disparity and multi-level parallelisms such as image line processing to deal with data dependency and irregular data access pattern problems in SGM. Later, to improve the robustness of SGM and achieve a more accurate stereo matching, Cambuim et al. [202] combine the sampling-insensitive absolute difference in the pre-processing phase, and propose a novel streaming architecture to detect noisy and occluded regions in the post-processing phase. The design is evaluated in a full stereo vision system using two heterogeneous platforms, DE2i-150 and DE4, and achieves a 25 FPS processing rate in 1024×768 HD maps with 256 disparity levels.

While most existing SGM designs on FPGA are implemented using the register-transfer level (RTL), some works leveraged the high-level synthesis (HLS) approach. Rahnama et al. [203] implement an SGM variation on FPGA using HLS, which achieves 72 FPS speed at 1242×375 pixel size with 128 disparity levels. To reduce the design effort and achieve an appropriate balance among speed, accuracy, and hardware cost, Zhao et al. [204] recently propose FP-Stereo for building high-performance SGM pipelines on FPGAs automatically. A series of optimization techniques are applied in this system to exploit parallelism and reduce resource consumption. Compared to GPU designs [205], it achieves the same accuracy at a competitive speed while consuming much less energy.

4.6   EFFICIENT LARGE-SCALE STEREO MATCHING ON FPGAS

4.6.1   ELAS ALGORITHM FRAMEWORK

Another popular stereo matching algorithm that offers a good trade-off between speed and accuracy is Efficient Large-Scale Stereo Matching (ELAS) [206]. ELAS is a generative probabilistic model using Bayesian approaches, which allows for dense matching with small aggregation windows by reducing ambiguities on the correspondences. It first establishes a prior over the disparity space by forming a triangulation on a set of correspondences whose estimation is simpler and comes with a higher degree of confidence. These correspondences provide a rough approximation of the scene geometry and guide the dense matching stage. ELAS is attractive since the

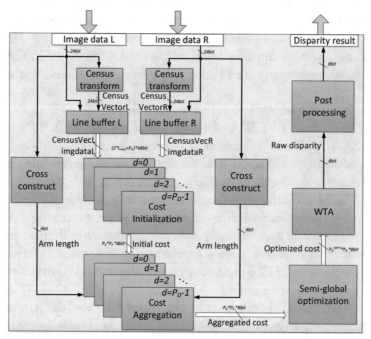

(a) Overall structure of the stereo matching module proposed in Wang et al. [198].

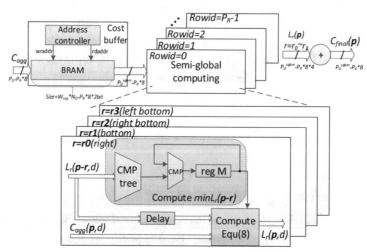

(b) Structure of the semiglobal optimization module of Wang et al. [198] design.

Figure 4.5: Semi-global stereo matching compute block diagram and its FPGA-based hardware architecture design. (Figures from [198].)

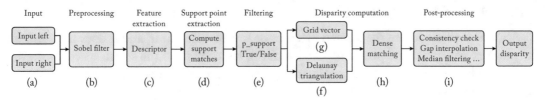

Figure 4.6: The overview of ELAS algorithm flow.

slanted plane prior can be very efficiently implemented, and the dense depth estimation is fully decomposable over all pixels.

Figure 4.6 demonstrates the overview of the ELAS algorithm. ELAS relies on that not all correspondences are equally difficult to be acquired. The detailed steps are described as follows.

Pre-Processing and Descriptor Extraction. The input stereo image pairs (left and right images) first pass through Sobel filters over the horizontal and vertical gradients to extract the descriptor information of each pixel (Fig. 4.6a–c).

Support Point Extraction. A set of sparse but confident correspondences (support points) is calculated using descriptor information. The support point set is obtained by comparing the distance between the first and second minima of the SAD evaluations across the disparity range. Only pixels with adequately unambiguous disparity values will be kept, so the resulting set is sparse (Fig. 4.6d).

Filtering. By comparing support point values to neighbors within a window region, two types of obtained support points are removed. The implausible values that are inconsistent would respectively corrupt the representation are filtered out. The redundant values that are identical to neighbors in the same row or column and unnecessarily complicate the coarse representation are also removed (Fig. 4.6e).

Disparity Computation. The filtered support points are then used to guide the dense stereo matching in two separate ways. First, they are used to construct a slanted plane prior for to coarse scene geometry approximation by conducting Delaunay triangulation. Second, they are aggregated and pooled within a sub-region to create grid vectors which makes stereo matching more reliable (Fig. 4.6f–h).

Post-Processing. ELAS uses several post-processing techniques, such as left-right consistency check, gap interpolation, median filtering, and small segment removal to invalidate occluded pixels and further smooth the images (Fig. 4.6i).

4.6.2   FPGA DESIGNS

ELAS is attractive because the slanted plane prior is piecewise linear, and it does not suffer in the presence of slanted and poorly textured surfaces, which reduces the search space. However, among the ELAS pipeline, the approximation of coarse scene geometry through triangulation of support points is a very iterative and sequential process with an unpredictable memory access pattern, impeding the acceleration of the whole pipeline.

To execute the ELAS more efficiently, Rahnama et al. [207] propose and implement an ARM + FPGA System-on-a-Chip (SoC), which achieves a frame rate of 47 FPS (up to 30× compared to high-end CPU) while consuming under 4 W of power. In this design, compute-intensive tasks, support point extraction, filtering, and dense matching, are offloaded to FPGA accelerators, whereas conditional and sequential tasks, Delaunay triangulation and grid vector extraction, are handled by ARM CPU. The combination of block RAMs and local memory is used to achieve higher parallelism and reduce data movement. The authors also reveal the strategy to accelerate complex and computationally diverse algorithms for low-power and real-time systems by collaboratively utilizing different compute components.

Rahnama et al. [208] propose a novel stereo approach that combines the best features of SGM and ELAS-based methods to compute highly accurate dense depth in real time. First, they use multiple passes of Fast R^3SFM (a fast SGM variant [203]), left-right consistency checking, and decimation to obtain a sparse but accurate set of disparity maps. Then, these correspondences are used as the support points for ELAS to obtain disparity priors from slanted planes. Finally, these disparity priors are incorporated into a final SGM-based optimization to produce robust dense predictions with high accuracy. This approach achieves an 8.7% error rate on the KITTI 2015 dataset at over 50 FPS, with a power consumption of only 4.5 W.

Gao et al. [209] further propose a fully FPGA-based architecture of ELAS system, *iELAS*, instead of offloading grid vector and Delaunay triangulation modules on Arm CPUs used in [207]. The original computational-intensive and irregular triangulation module is re-formed in a regular manner with points interpolation in *iELAS*, which is more hardware-friendly. The *iELAS* algorithm flow is shown in Fig. 4.7. After extracting support points, an interpolation module is added to derive a set of newly support points with fixed numbers and co-ordinations, which facilitates the construction of slanted planes. The interpolation is performed using the disparity value of the support points within the neighborhood to fill the vacant positions. The detailed steps are as follows.

Horizontal Interpolation. For the position s to be interpolated, first search the support points in the horizontal direction within $(s - s_\delta, s + s_\delta)$ window. If there are support points (P_L, P_R) lying on both sides and their disparity values $|D_{P_L} - D_{P_R}| \leq \epsilon$, then we use the mean of (D_{P_L}, D_{P_R}) to interpolate. If $|D_{P_L} - D_{P_R}| > \epsilon$, then $min(D_{P_L}, D_{P_R})$ will be chosen for interpolation.

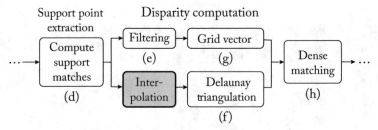

(a) The interpolated ELAS algorithm flow, the original
ELAS algorithm flow is shown in Fig. 4.6.

36			38				38
		26		38			
38							
			46			32	
		24					
	54					54	
			46				
	32					52	

Interpolation →

36	37	37	38	38	38	38	38
37	0	26	26	38	0	0	0
38	0	25	26	0	0	0	0
0	0	25	46	32	32	32	0
0	0	24	46	0	0	32	0
0	54	54	54	54	54	54	0
0	32	0	46	0	0	53	0
0	32	32	32	32	32	52	0

(b) Exampled support points interpolation (s_δ = 5, ϵ = 3, C = 0).
Red represents horizontal interpolation, blue represents vertical
interpolation, and green represents constant interpolation.

Figure 4.7: The overview and an example case of interpolated ELAS (*iELAS*) implementation
proposed in [209].

Vertical Interpolation. If no support point pair (P_L, P_R) is found in the horizontal direction, then search in the vertical direction to find (P_T, P_B) and perform interpolation using the same method as step 1.

Constant Interpolation. If no support point pairs are found in both horizontal and vertical directions, then fill a constant disparity value C in the position s.

Following this procedure, we present an example of support points interpolation in Fig. 4.7b, where $s_\delta = 5$, $\epsilon = 3$, $C = 0$.

The evaluation results demonstrate *iELAS* achieves up to 3.32× and 38.4× speedup in frame rate, and 1.13× and 27.1× improvement in energy efficiency when compared to the Arm+FPGA [207] and Intel i7 CPU, respectively.

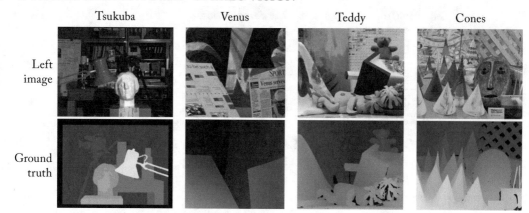

Figure 4.8: Standard benchmarking datasets from the Middlebury. (Figure adapted from [213].)

4.7   EVALUATION AND DISCUSSION

4.7.1   DATASET AND ACCURACY

Datasets that provide a sufficient number of samples for specific problems have proven to be an essential catalyst for rapid solution improvement in many fields. Based on the quantitative evaluation, they facilitate fast iteration of algorithms, expose potential weakness, and enable a fair comparison. In the stereo vision system, two commonly used datasets are Middlebury [210] and KITTI [211].

Middlebury Dataset

Middlebury dataset was developed by Microsoft Research in 2001. The goals of this project are two-fold: (1) provide a taxonomy of existing stereo algorithms that allows the dissection and comparison of individual algorithm components design decisions; and (2) provide a test bed for the quantitative evaluation of stereo algorithms. Figure 4.8 presents exampled benchmark images from Middlebury dataset. More details can be found in [210, 212].

KITTI Dataset

KITTI dataset is a joint project between Karlsruhe Institute of Technology and Toyota Technological Institute at Chicago in 2012. The purpose of this project is to collect a realistic and challenging dataset for autonomous driving. The complete KITTI dataset consists of stereo and optical flow data, visual odometry data, object detection and orientation data, object tracking data, and road parsing data. KITTI dataset consists of 200 training scenes and 200 test scenes. More details can be found in [211].

Evaluation

The depth quality of some FPGA implementations mentioned above is evaluated on Middlebury Benchmark, with four image pairs: Tsukuba, Venus, Teddy, and Cones (Fig. 4.8). The results are summarized in Table 4.1, and it is shown that there is a general trade-off between accuracy and processing speed.

The adopted evaluation metric is "Percentage of bad matching pixels" ("Average error" column in Table 4.1). It can be calculated based on the known ground truth map $d_T(x, y)$ and the computed depth map $d_C(x, y)$:

$$B = \frac{1}{N} \sum_{(x,y)} (|d_C(x, y) - d_T(x, y)| > \delta_d),$$

where N is the total number of pixels and δ_d is a disparity error tolerance.

4.7.2   POWER AND PERFORMANCE

Table 4.2 summarizes the performance, power, and resource utilization of some stereo vision FPGA designs based on their targeted algorithms. The processing speed of these systems is described in FPS, and more meaningfully, in a million disparity estimations per second (MDE/s = height × width × disparity × FPS).

Then the stereo vision system designs in Table 4.2 are drawn as points in Fig. 4.9 (if both power and speed number are reported). We use $\log_{10}(power)$ as x-coordinate and $\log_{10}(speed)$ as y-coordinate ($y - x = \log_{10}(energy\_efficiency)$). Besides FPGA-based implementations, we also plot GPU and CPU experimental results as a comparison to FPGA designs' performance. In general, local and semi-global stereo matching designs have achieved higher performance and energy efficiency than global stereo matching designs. As introduced in Section 4.4, global stereo matching algorithms usually involve massive computational-intensive optimization techniques. Even for the same design, varying design parameters (e.g., window size) may result in a 10× difference in energy efficiency. Compared to GPU and CPU-based designs, FPGA-based designs have achieved higher energy efficiency, and the speed of many FPGA implementations has surpassed general-purpose processors.

4.8   SUMMARY

In this chapter, we have reviewed various stereo vision algorithms in robotic perception and their FPGA accelerator designs. The whole computation pipeline is analyzed. Several design techniques have been summarized, which are used to accelerate the computational bottleneck based on characterization results. According to the evaluation results, with careful algorithm-hardware co-design, FPGAs can achieve two orders of magnitude of higher energy efficiency and performance than the state-of-the-art GPUs and CPUs. This illustrates that FPGAs are a promising candidate for accelerating stereo vision systems.

Table 4.1: A comparison between different designs on performance (MDE/s) and accuracy results on Middlebury Benchmark, where MDE/s = width × height × FPS × disparity. (The lower of average bad pixel rate (error) means the better stereo matching performance.)

Design	MDE/s	Tsukuba			Venus			Teddy			Cones			Average Error
		nonocc[1]	all[2]	disc[3]	nonocc	all	disc	nonocc	all	disc	nonocc	all	disc	
[214]	15437	-	24.5	-	-	15.7	-	-	15.1	-	-	14.1	-	all = 17.3
[215]	13076	3.62	4.15	14.0	0.48	0.87	2.79	7.54	14.7	19.4	3.51	11.1	9.64	7.65
[198]	10472	2.39	3.27	8.87	0.38	0.89	1.92	6.08	12.1	15.4	2.12	7.74	6.19	5.61
[188]	9362	1.66	2.17	7.64	0.4	0.6	1.95	6.79	12.4	17.1	3.34	8.97	9.62	6.05
[185]	4522	9.79	11.6	20.3	3.59	5.27	36.8	12.5	21.5	30.6	7.34	17.6	21.0	17.2
[186]	3020	3.84	4.34	14.2	1.2	1.68	5.62	7.17	12.6	17.4	5.41	11.0	13.9	8.2
[200]	1455	4.1	-	-	2.7	-	-	11.4	-	-	8.4	-	-	nonocc = 6.7
[196]	590	1.43	2.51	6.6	2.37	2.97	13.1	8.11	13.6	15.5	8.12	13.8	16.4	8.71

[1] nonocc: average percentage of bad pixels in non-occluded regions.
[2] all: average percentage of bad pixels in all regions.
[3] disc: average percentage of bad pixels in discontinuous regions.

Table 4.2: Comparison of Stereo Vision Systems on FPGA platforms, across local stereo matching, global stereo matching, semi-global stereo matching (SGM), and efficient large-scale stereo matching (ELAS) algorithms. The results reported in each design are evaluated by frame rate (FPS), image resolution (width × height), disparity levels, million disparity estimations per second (MDE/s), power (W), resource utilization (logic% and BRAM%), and hardware platforms.

Algorithm	Reference	Frame Rate (FPS)	Image Resolution (Width × Height)	Disparity Level	MDE/s	Power (W)	Resource(%) Logic/ BRAM
Local stereo matching	Jin et al. [185]	230	640 × 480	64	4522	–	34.0 / 95.0
	Zhang et al. [186]	60	1024 × 768	64	3020	1.56	61.8 / 67.0
	Honegger et al. [187]	127	376 × 240	32	367	2.8	49.0 / 68.0
	Jin et al. [188]	507.9	640 × 480	60	9362	3.35	81.0 / 39.7
Global stereo matching	Park et al. [194]	30	320 × 240	128	295	–	– / –
	Sabihuddin et al. [195]	63.54	640 × 480	128	2498	–	23.0 / 58.0
	Jin et al. [196]	32	640 × 480	60	590	1.40	72.0 / 46.0
	Zha et al. [172]	30	1920 × 1680	60	5806	–	84.8 / 91.9
	Puglia et al. [173]	30	1024 × 768	64	1510	0.17	57.0 / 53.0
Semi-global stereo matching	Banz et al. [200]	37	640 × 480	128	1455	2.31	51.2 / 43.2
	Wang et al. [198]	42.61	1600 × 1200	128	10472	2.79	93.9 / 97.3
	Cambuim et al. [201]	127	1024 × 768	128	12784	–	– / –
	Rahnama et al. [203]	72	1242 × 375	128	4292	3.94	75.7 / 30.7
	Cambuim et al. [202]	25	1024 × 768	256	5033	6.5	50.0 / 38.0
	Zhao et al. [204]	147	1242 × 375	64	4382	9.8	68.7 / 38.7
ELAS	Rahnama et al. [207]	47	1242 × 375	–	–	2.91	11.9 / 15.7
	Rahnama et al. [208]	50	1242 × 375	–	–	5	70.7 / 8.7

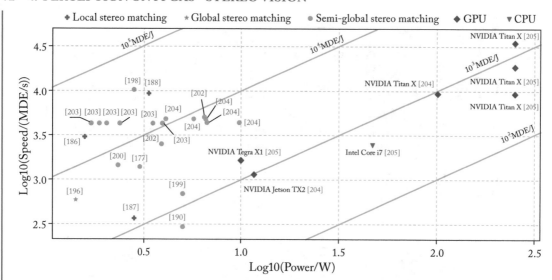

Figure 4.9: A comparison between different designs for perception tasks on a logarithm coordinate of power (W) and performance (MDE/s).

Two potential aspects can be further optimized. On the one hand, most current designs accelerate key computationally intensive stereo vision kernels, but do not take the end-to-end system impact into account. For example, image data preprocessing and movement usually take much compute time in the whole processing pipeline and will impact the performance of perception. Near-sensor compute is one promising way to improve on that. On the other hand, FPGAs are usually constrained by their onboard resources. Therefore, how to intelligently manage memory resources, dataflow, and interconnect network, to make FPGA accelerators support large-scale and high-resolution images required to be considered in the future design.

CHAPTER 5

Localization on FPGAs

Localization, i.e., ego-motion estimation, is one of the most fundamental tasks to autonomous machines, in which an agent calculates the position and orientation of itself in a given frame of reference, i.e., map. For general robotic software stacks, localization is the building block of many tasks. Knowing the translational pose enables a robot or a self-driving car to plan the path and to navigate. The rotational pose lets robots and drones stabilize themselves [216, 217].

Localization algorithms estimate poses incrementally from one or more onboard sensors, such as camera, LiDAR, IMU, and GPS. To achieve high accuracy, huge computing resources are needed to process the large volume of sensory data in real-time. This poses a critical challenge on the localization systems of autonomous robots, wherein power budget and thermal and physical size restrictions are stringent.

Compared with CPUs and GPUs, FPGA is more suitable for robotic localization systems. Modern FPGAs have rich and re-configurable sensor interfaces, which are able to provide well-synchronized sensory data from various sensors. FPGAs have massive re-configurable fabrics and dedicated hardware blocks, which enables accelerator designs for complex localization algorithms.

While literature is rich with accelerator designs for specific localization algorithms [71, 218–221], prior efforts are mostly pointing solutions in that they each specialize for a particular localization algorithm. However, each algorithm suits only a particular operating scenario, whereas autonomous machines routinely operate under different scenarios.

This chapter presents a unified FPGA-based framework for localization that adapts to different operating scenarios. The localization framework adapts to different operating scenarios by unifying core primitives in existing algorithms, and thus provides a desirable software baseline to identify general acceleration opportunities. The framework exploits the inherent two-phase design of existing localization algorithms, which share a similar visual frontend but have different optimization backends (e.g., probabilistic estimation vs. constrained optimization) that suit different operating scenarios.

5.1 PRELIMINARY

5.1.1 CONTEXT

Formally, localization is the task that generates the six degrees of freedom (DoF) pose in a reference frame, as shown in Fig. 5.1. The six DoF includes the three DoF for the translational

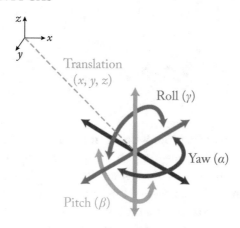

Figure 5.1: Localization estimates the six degrees of freedom (6 DoF) pose of a body $(x, y, z, \alpha, \beta, \gamma)$.

pose, which specifies the $< x, y, z >$ position, and the three DoF for the rotational pose, which specifies the orientation about three perpendicular axes, i.e., yaw, roll, and pitch.

Sensing   Localization is made possible through sensors that interact with the environment. Common sensors include cameras, LiDARs, Inertial Measurement Units (IMU), and Global Positioning System (GPS) receivers.

- GPS: GPS receivers directly provide the three translational DoF in a global coordinate. However, they are not used alone because their signals (1) do not provide the three rotational DoF, (2) are blocked in an indoor environment, and (3) could be degenerate when the multi-path problem occurs [222].

- IMU: An IMU provides the relative 6 DoF information by combining a gyroscope and an accelerometer. IMU samples are noisy [223]. Localization results would quickly drift if relying completely on the IMU. Thus, IMU is usually combined with other sensors in localization.

- LiDAR or Camera: Given a map composed of LiDAR or Camera measurements of landmarks, poses could be deduced by comparing the sensory observation with the landmark map. For scenarios without a prior map, localization is achieved by solving the simultaneous localization and mapping (SLAM) problem. Given sensory observations and robot motion information, SLAM algorithms jointly estimate the sensory representation of the 3D environment (i.e., sensory map) and poses of the robot.

Operating Environment   Autonomous machines usually have to operate under different scenarios to finish a task. For instance, logistics robots that transfer cargo between warehouses in

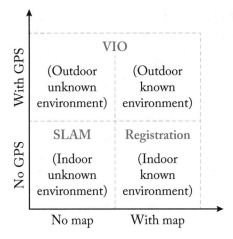

Figure 5.2: Taxonomy of real-world environments. Each environment prefers a particular localization algorithm.

an industrial park may navigate both outside and inside warehouses with a prior map. When the robots are moved to a different section of the park, the robots would spend time on mapping new warehouses for further operation.

In principle, real-world environments can be classified along two dimensions for localization purposes as shown in Fig. 5.2: the availability of a pre-constructed map and the availability of a direct localization sensor, mainly the GPS. First, an environment's map could be available depending on whether the environment has been previously mapped, i.e., known to the autonomous machine. A map could significantly ease localization because the map provides a reliable frame of reference. Second, depending on whether the environment is indoor or outdoor, GPS could provide absolute positioning information that greatly simplifies localization. Overall, four scenarios exist in the real world: (1) indoor unknown environment; (2) indoor known environment; (3) outdoor unknown environment; and (4) outdoor known environment.

5.1.2   ALGORITHM OVERVIEW

To understand the most fitting algorithm in each scenario, three fundamental categories of localization algorithms [216, 224] are analyzed in this chapter, which are: (1) complementary to each other in requirements and capabilities, (2) well-optimized by algorithmic researchers, and (3) widely used in industry. LiDAR-based and camera-based localization algorithms are similar in algorithm frameworks. Without losing generality, this section focuses on vision-based algorithms, which are the following.

- **Registration**: Given camera observations and a landmark map, map registration calculates 6 DoF poses in the map coordinate that best aligns the observations with the landmarks. The "bag-of-words" framework [225–227] is considered in this chapter, which is the backbone of many products such as iRobot [228].

- **VIO**: A classic way of localization without an explicit global map is to calculate relative poses incrementally from consecutive sensor measurements, such as images or lidar point clouds. Probabilistic nonlinear filters are effective to solve this kind of problem. Kalman Filter (KF) is the most practical and effective nonlinear filter for pose estimation, in which relative poses of an autonomous machine are calculated from a period of sensor observations and motion commands. Common KF extensions include Extended Kalman Filter (EKF) [229] and Multi-State Constraint Kalman Filter (MSCKF) [230]. Since the state-of-the-art KF-based algorithms fuse camera observations and IMU data [231–233], we simply refer to them as VIO (visual-inertial odometry).

  In this section, we use an MSCKF-based framework [233, 234] due to its superior accuracy compared to other state estimation algorithms. For instance on the EuRoC dataset, MSCKF accuracy on average 0.18 m error reduction over EKF [233]. MSCKF is also used by many products such as Google ARCore [235].

  Since VIO calculates the relative trajectory using local observations, its localization errors could accumulate over time [236]. One effective mitigation is to provide VIO with absolute positioning information through GPS. When stably available, GPS signals help the VIO algorithm relocalize to correct the localization drift [237]. VIO coupled with GPS is often used for outdoor navigation such as in DJI drones [238].

- **SLAM**: It simultaneously constructs a map while localizing an agent within the map. SLAM reduces the accumulated errors in VIO by constantly remapping the global environment and closing the loop. SLAM is usually formulated as a constrained optimization problem, often through bundle adjustment [239]. SLAM algorithms are used in many robots [240] and AR devices (e.g., Hololens) [241]. The widely used VINS-Fusion framework [242, 243] is used in this section.

The design space of VIO and SLAM is broader than the specific algorithms we target. For instance, VIO could use factor-graph optimizations [219] rather than KF. We use VIO to refer to algorithms using probabilistic state estimation without explicitly constructing a map, and uses SLAM to refer to optimization-based algorithms that explicitly construct a map. It is these fundamental classes of algorithms that we focus on.

Accuracy Characterizations   Figures 5.3a–5.3d show the root-mean-square error (RMSE) (y-axis) and the average performance (x-axis) of the three localization algorithms under the four scenarios, respectively. They demonstrate that there is no single localization algorithm that fits all scenarios. Instead, each operating environment tends to prefer a different localization algorithm to minimize error.

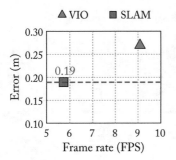

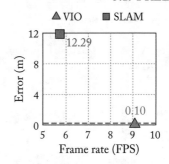

(a) Indoor unknown environment. SLAM provides the best accuracy.

(b) Outdoor unknown environment. VIO is best in accuracy.

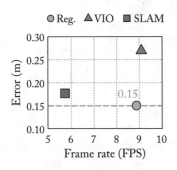

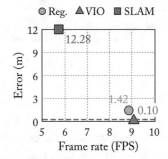

(c) Indoor known environment. Registration has the best accuracy.

(d) Outdoor known environment. VIO is best in accuracy.

Figure 5.3: Localization error vs. performance trade-off in the four operating scenarios, each requiring a particular localization algorithm to maximize accuracy. Registration does not apply to scenarios without a map. The KITTI Odometry dataset [244] are used. The performance data is collected from a four-core Intel Kaby Lake CPU.

In an indoor environment without a map, Fig. 5.3a shows that SLAM delivers a much lower error than VIO. The registration algorithm is not applicable in this case as it requires a map. We do not supply the GPS signal to the VIO algorithm here due to the unstable signal reception; supplying unstable GPS signals would worsen the VIO accuracy. The data shows that VIO lacks the relocalization ability to correct accumulated drifts without GPS.

When an indoor environment has a pre-constructed map, registration achieves higher accuracy while operating at a higher frame rate than SLAM as shown in Fig. 5.3c. Specifically, the registration algorithm has only a 0.15 m localization error while operating at 8.9 FPS. VIO almost doubles the error (0.27 m) due to drifts, albeit operating at a slightly higher frame rate (9.1 FPS).

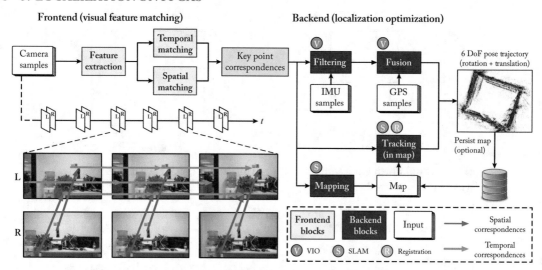

Figure 5.4: Overview of the unified localization algorithm framework that adapts to different operating environments. The framework consists of a vision frontend and an optimization backend. The frontend establishes feature correspondences and is always activated. The backend operates under three modes, each activated in a particular operating scenario. Essentially, our localization framework fuses the three primitive localization algorithms (i.e., registration, VIO, and SLAM) by sharing common building blocks. This strategy preserves the accuracy under each scenario while minimizing algorithm-level redundancies.

VIO becomes the best algorithm outdoor, both with (Fig. 5.3d) or without a map (Fig. 5.3b). VIO achieves the highest accuracy with the help of GPS (0.10 m error) and is the fastest. Even with a pre-constructed map (Fig. 5.3d), registration still has a higher error (1.42 m) than VIO due to measurement noises and mismatching. SLAM is the slowest and has a significantly higher error due to the lack of GPS signals or a prior map for accurate relocation.

Figure 5.2 summarizes the algorithm affinity of each operating scenario to maximize accuracy: indoor environment with a map prefers registration; indoor environment without a map prefers SLAM; outdoor environment prefers VIO (with GPS corrections). As an autonomous machine often operates under different scenarios, a localization system must simultaneously support the three algorithms so as to be useful in practice.

5.2   ALGORITHM FRAMEWORK

This section presents a vision-based localization framework that flexibly adapts to different operating environments. Figure 5.4 shows the algorithmic framework. This framework captures general patterns and shares common building blocks across the three primitive algorithms (i.e., registration, VIO, and SLAM). In particular, the three primitive algorithms share a two-phase pipeline, which consists of a visual feature matching phase and a localization optimization phase. While the optimization technique differs in the three primitive algorithms, the feature matching phase is the same.

Decoupled Framework   The framework consists of a shared vision frontend, which extracts and matches visual features and is always activated, and an optimization backend, which has three modes—registration, VIO, and SLAM—each triggered under a particular operating scenario (Fig. 5.2). Each mode forms a unique dataflow path by activating a set of blocks in the backend, as shown in Fig. 5.4.

Sharing the frontend across different backend modes is driven not only by the algorithmic characteristics but also by the hardware resource efficiency. Vision frontend is resource-heavy in localization accelerators. For instance, the frontend area contributes to about 27% and 53% of the total chip area in two recent ASIC accelerators [218, 219]. The same is true in FPGA implementations [71, 220].

Below we describe each block in the framework. Each block is directly taken from the three individual algorithms, which have been well-optimized and used in many products. While some components have been individually accelerated before [245, 246], it is not known how to provide a unified architecture to efficiently support these components in one system, which is the goal of our hardware design.

Frontend   The visual frontend extracts visual features to find correspondences in consecutive observations, both temporally and spatially, which the backend uses to estimate the pose. In particular, the frontend consists of three blocks.

- **Feature Extraction** The frontend first extracts key feature points, which correspond to salient landmarks in the 3D world. Operating on feature points, as opposed to all image pixels, improves the robustness and compute efficiency of localization. In particular, key points are detected using the widely used FAST feature [247]; each feature point is associated with an ORB descriptor [61] in preparation for spatial matching later.

- **Stereo Matching** This block matches key points spatially between a pair of stereo images (from the left and right cameras) to recover depth information. Widely used blocking matching method [248] based on the ORB descriptor of each feature point is used as an example [61, 249].

- **Temporal Matching** This block establishes temporal correspondences by matching the key points of two consecutive images. Instead of searching for matches, this block tracks feature points across frames using the classic Lucas-Kanade optical flow method [62].

Backend   The backend calculates the 6 DoF pose from the visual correspondences generated in the frontend. Depending on the operating environment, the backend is dynamically configured to execute in one of the three modes, essentially providing the corresponding primitive algorithm. To that end, the backend consists of the following blocks.

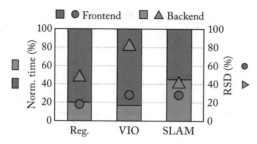

Figure 5.5: Latency distribution and relative standard deviation (RSD) of frontend and backend in three modes.

- **Filtering** This block is activated only in the VIO mode. It uses Kalman Filter to integrate a series of measurements observed over time, including the feature correspondences from the frontend and the IMU samples, to estimate the pose. We use MSCKF [230], a Kalman Filter framework that keeps a sliding window of past observations rather than just the most recent past.

- **Fusion** This block is activated only in the VIO mode. It fuses the GPS signals with the pose information generated from the filtering block, essentially correcting the cumulative drift introduced in filtering. We use a loosely coupled approach [232], where the GPS positions are integrated through a simple EKF [229].

- **Mapping** This block is activated only in the SLAM mode. It uses the feature correspondences from the frontend along with the IMU measurements to calculate the pose and the 3D map. This is done by solving a nonlinear optimization problem, which minimizes the projection errors (imposed by the pose estimation) from 2D features to 3D points in the map. The optimization problem is solved using the Levenberg-Marquardt (LM) method [250]. We target an LM implementation in the Ceres Solver, which is used in products such as Google's Street View [251]. In the end, the generated map could be optionally persisted offline and later used in the registration mode.

- **Tracking** This block is activated both in the registration and the SLAM mode. Using the bag-of-words place recognition method [225, 226], this block estimates the pose based on the features in the current frame and a given map. In the registration mode, the map is naturally provided as the input. In the SLAM mode, tracking and mapping are executed in parallel, where the tracking block uses the latest map generated from the mapping block, which continuously updates the map.

Latency Distribution Figure 5.5 shows the average latency distribution between the frontend and the backend across the three scenarios. It shows that the frontend contributes significantly

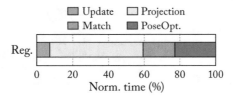

Figure 5.6: Latency distribution in registration backend.

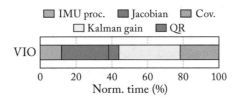

Figure 5.7: Latency distribution in VIO backend.

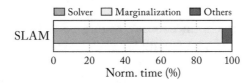

Figure 5.8: Latency distribution in SLAM backend.

to the end-to-end latency across all three scenarios. Since the frontend is shared across different backend modes, accelerating the frontend would lead to "universal" performance improvement. The frontend time varies from 55% in the SLAM mode to 83% in the VIO mode. The SLAM backend is the heaviest because it iteratively solves a complex nonlinear optimization problem.

While the frontend is a lucrative acceleration target, backend time is non-trivial too, especially after the frontend is accelerated. Figures 5.6, 5.7, and 5.8 show the distribution of different blocks within each backend mode. Marginalization in SLAM, computing Kalman gain in VIO, and projection in registration are the biggest contributors to the three backend modes. These three kernels also contribute significantly to the backend variations.

5.3   FRONTEND FPGA DESIGN

5.3.1   OVERVIEW

Figure 5.9 provides an overview of the frontend architecture. The input (left and right) images are streamed through the DMA and double-buffered on-chip. The on-chip memory design

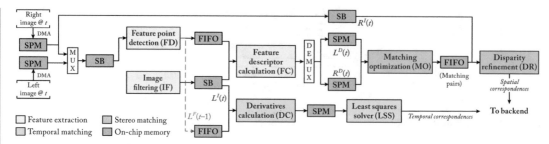

Figure 5.9: Frontend architecture, which consists of three blocks: feature extraction, optical flow, and stereo matching. The left and right camera streams are time-multiplexed in the feature extraction block to save hardware resources. On-chip memories are customized in different ways to suit different types of data reuse: stencil buffer (SB) supports stencil operations, FIFO captures sequential accesses, and scratchpad memory (SPM) supports irregular accesses.

mostly allows the frontend to access DRAM only at the beginning and the end of the pipeline as we will discuss later.

The two images go through three blocks: feature extraction, spatial matching, and temporal matching. Each block consists of multiple tasks. For instance, the feature extraction block consists of three tasks: feature point detection, image filtering, and descriptor calculation.

The feature extraction block is exercised by both the left and right images. The feature points in the left image at time $t - 1$ (L_{t-1}^F) are buffered on-chip, which are combined with the left image at time t (L_t^I) to calculate the temporal correspondence at t. Meanwhile, the feature descriptors in both images at t (L_t^D and R_t^D) are consumed by the stereo matching block to calculate the spatial correspondences at t. The temporal and spatial correspondences are about 2–3 KB on average; they are transmitted to the backend.

5.3.2 EXPLOITING TASK-LEVEL PARALLELISMS

Understanding the Parallelisms Figure 5.10 shows the detailed task-level dependencies in the frontend. At the high level, feature extraction (FE) consumes both the left and right images, which are independent and could be executed in parallel. Stereo matching (SM) must wait until both images finish the Feature Descriptor Calculation (FC) task in the FE block because SM requires the feature points/descriptors generated from both images. Temporal matching (TM) operates only on the left image and is independent of SM. Thus, TM could start whenever the left image finishes the image filtering (IF) task in FE. The IF task and the feature point detection (FD) task operate in parallel in FE.

TM consists of two serialized tasks: derivatives calculation (DC), whose outputs drive a (linear) least squares solver (LSS). SM consists of a matching optimization (MO) task, which provides initial spatial correspondence by comparing hamming distances between feature de-

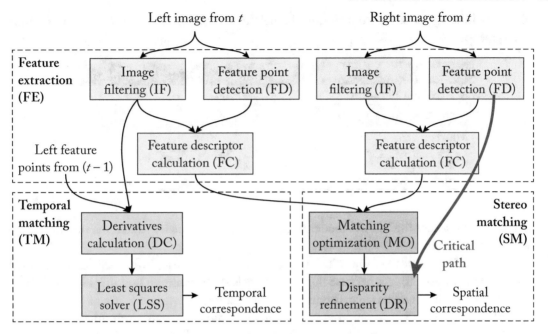

Figure 5.10: Task dependencies in the frontend.

scriptors, and a disparity refinement (DR) task which refines the initial correspondences through block matching.

Design TM latency is usually over 10× lower than SM latency. Thus, the critical path is FD → FC → MO → DR. The critical path latency is in turn dictated by the SM latency (MO+DR), which is roughly 2–3× higher than the FE latency (FD+FC). Therefore, we pipeline the critical path between the FE and the SM. Pipelining improves the frontend throughput, which is dictated by the latency of SM.

The FE hardware resource consumption roughly doubles the resource consumption of TM and SM combined, as FE processes raw input images whereas SM and TM process only key points ($< 0.1\%$ of total pixels). Thus, we time-share the FE hardware between the left and right images. This reduces hardware resource consumption, but does not hurt the throughput since FE is much faster than SM.

5.4 BACKEND FPGA DESIGN

Building Blocks Figures 5.6–5.8 show that each backend mode inherently possesses a kernel that contributes significantly to both the overall latency and the latency variation: camera model projection under the registration mode, computing Kalman gain under the VIO mode,

Table 5.1: Latency variation-contributing kernels in the backend are composed of common matrix operations

Building Block	Projection	Kalman Gain	Marginalization
Matrix multiplication	✓	✓	✓
Matrix decomposition		✓	✓
Matrix inverse			✓
Matrix transpose		✓	✓
Fwd./bwd. substitution		✓	✓

marginalization under the SLAM mode. Accelerating these kernels reduces the overall latency and latency variation.

While it is possible to spatially instantiate separate hardware logic for each kernel, it would lead to resource waste. This is because the three kernels share common building blocks. Fundamentally, each kernel performs matrix operations that manipulate various forms of visual features and IMU states. Table 5.1 decomposes each kernel into different matrix primitives.

For instance, the projection kernel in the registration mode simply multiplies a 3×4 camera matrix $\mathbf{C}$ with a $4 \times M$ matrix $\mathbf{X}$, where M denotes the number of feature points in the map (each represented by 4D homogeneous coordinates). The Kalman gain $\mathbf{K}$ in VIO is computed by:

$$\mathbf{S} = \mathbf{H} \times \mathbf{P} \times \mathbf{H}^T + \mathbf{R} \tag{5.1a}$$
$$\mathbf{S} \times \mathbf{K} = \mathbf{P} \times \mathbf{H}^T, \tag{5.1b}$$

where $\mathbf{H}$ is the Jacobian matrix of the function that maps the true state space into the observed space, $\mathbf{P}$ is the covariance matrix, and $\mathbf{R}$ is an identity noise matrix. Calculating $\mathbf{K}$ requires solving a system of linear equations, which is implemented by matrix ($\mathbf{S}$) decomposition followed by forward/back-substitution. Marginalization combines all five operations [243].

Design The backend design specializes the hardware for the five matrix operations in Table 5.1, which the three backend kernels are then mapped to. These matrix operations are low-level enough to allow sharing across different backend kernels but high level enough to reduce the control flows, which are particularly inefficient to implement on FPGAs.

Figure 5.11 shows the backend architecture. The input and output are buffered on-chip and DMA-ed from/to the host. The inputs to each matrix block are stored in the scratchpad memories (SPMs). The input matrices must be ready before an operation starts. The architecture accommodates different matrix sizes by exploiting the inherent blocking nature of matrix operations (e.g., multiplication, decomposition), where the output could be computed by iteratively operating on different blocks of the input matrices. Thus, the compute units have to support

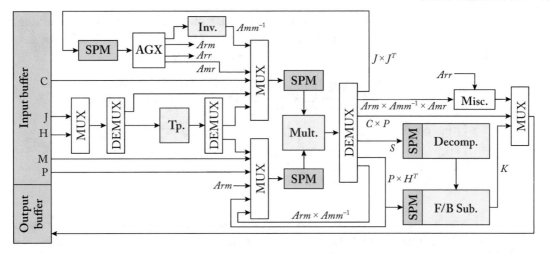

Figure 5.11: Backend architecture. It consists of five basic building blocks for matrix operations: inverse (Inv.), decomposition (Decomp.), transpose (Tp.), multiplication (Mult.), and forward/backward-substitution. "AGX" is the address generation logic, and "Misc." contains the rest of the logic not included in any above (e.g., addition). Scratchpad memories (SPM) are used to stored intermediate data. The input and output is buffered on-chip and DMA-ed from/to the host.

computations for only a block, although the SPMs need to accommodate the size of the entire input matrices.

Optimization Unique characteristics inherent in various matrices can be exploited to further optimize the computation and memory usage. The **S** matrix is inherently symmetric (Eq. (5.1a)). Thus, the computation and storage cost of **S** can naturally be reduced by half. In addition, the matrix that requires inversion in marginalization, A_{mm}, is a symmetric matrix with a unique blocking structure of $\begin{bmatrix} A & B \\ C & D \end{bmatrix}$, where A is a diagonal matrix and D is a 6×6 matrix, where 6 represents the number of degrees of freedom in a pose to be calculated. Therefore, the inversion hardware is specialized for a 6×6 matrix inversion combined with simple reciprocal structures.

5.5 EVALUATION

5.5.1 EXPERIMENTAL SETUP

Hardware Platform To show the flexibility and general applicability of the localization framework, FPGA platforms targeting autonomous vehicles and drones, which represent two ends of the spectrum of autonomous machines, are designed. The Car FPGA platform is instantiated on a Xilinx Virtex-7 XC7V690T FPGA board [252] connected to a PC machine, which has a four-core Intel Kaby Lake CPU. The Drone platform is built on a Zynq Ultrascale+ ZU9CG

Table 5.2: FPGA resource consumption of Car and Drone and their utilizations on the actual FPGA boards. The "N.S." columns denote the hypothetical resource usage if we do not share the frontend and the building blocks in the backend across the three modes.

Resource	Car	Virtex-7	N.S.	Drone	Zynq	N.S.
LUT	350671	80.9%	795604	231547	84.5%	659485
Flip-Flop	239347	27.6%	628346	171314	31.2%	459485
DSP	1284	35.6%	3628	1072	42.5%	3064
BRAM	5.0	87.5%	13.2	3.67	92.3%	10.6

board [253], which integrates a quad-core ARM Cortex-A53 CPU with an FPGA on the same chip. The actual accelerator implementations on both instances are almost the same except that the Car design uses a larger matrix multiplication/decomposition unit and larger line buffers and scratchpad memories to deal with a higher image resolution.

The FPGA is directly interfaced with the cameras and IMU/GPS sensors. The host and the FPGA accelerator communicate three times in each frame: the first time from the FPGA, which transfers the frontend results and the IMU/GPS samples to the host; the second time from the host, which passes the inputs of a backend kernel (e.g., the **H**, **P**, and **R** matrices to calculate Kalman gains) to the FPGA, and the last time transferring the backend results back to the host.

Baselines Today's localization systems are mostly implemented on general-purpose CPU platforms. Thus, the Car design is compared against the software implementation on the PC machine without the FPGA. The Drone design is compared against the software implementation on the quad-core Arm Cortex-A57 processor on the TX1 platform [254], a representative mobile platform. To obtain stronger baselines, the software implementations leverage multi-core and the SIMD capabilities of the CPUs.

Dataset For the Drone design, EuRoC [255] (Machine Hall sequences) is used, which is a widely used drone dataset. Since EuRoC contains only indoor scenes, we complement EuRoC with our in-house outdoor dataset. The input images are sized to the 640×480 resolution. For the Car design, KITTI Odometry [244] (grayscale sequence) is used, which is a widely used self-driving car dataset. Similarly, since KITTI contains only outdoor scenes, we complement it with our in-house dataset for indoor scenes. The input images are uniformly sized to the 1280×720 resolution.

5.5.2 RESOURCE CONSUMPTION

Table 5.2 lists the FPGA resource usage. Overall, the Car instance consumes more resources than the Drone as the former uses larger hardware structures to cope with higher input resolu-

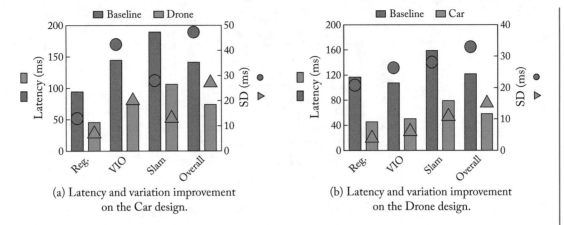

(a) Latency and variation improvement on the Car design.

(b) Latency and variation improvement on the Drone design.

Figure 5.12: Overall system latency and variation reduction.

tions. To demonstrate the effectiveness of our hardware design that shares the frontend and the various backend building blocks across the three modes (Table 5.1), the "N.S." columns show the hypothetical resource consumption without sharing these structures in both instances. Resource consumption of all types would more than double, exceeding the available resources on the FPGA boards.

Frontend dominates resource consumption. In Car, the frontend uses 83.2% LUT, 62.2% Flip-Flop, 80.2% DSP, and 73.5% BRAM of the total used resource; the percentages in Drone are similar. In particular, feature extraction consumes over two-thirds of frontend resources, corroborating our design decision to multiplex the feature extraction hardware between left and right camera streams (Section 5.3.2).

5.5.3   PERFORMANCE

Car Results   Figure 5.12a compares the average frame latency and standard deviation (SD) of latency between the baseline and the Car FPGA. It shows overall results as well as the results in the three modes separately. The end-to-end frame latency is reduced by 2.5×, 2.1×, and 2.0× in registration, VIO, and SLAM mode, respectively, which leads to an overall 2.1× speedup. The FPGA design also significantly reduces the variation. The SD is reduced by 58.4%. The latency reduction directly translates to higher throughput, which reaches 17.2 FPS from 8.6 FPS, as shown in Fig. 5.13. Further pipelining the frontend with the backend improves the FPS to 31.9.

The energy is also significantly reduced. Figure 5.14 compares the energy per frame between the baseline and the Car FPGA. With hardware acceleration, the Car FPGA reduces the average energy by 73.7%, from 1.9–0.5 J per frame.

Drone Results   Figure 5.12b compares the average frame latency and variation between the baseline and the Drone FPGA. The frame latency has a speedup of 2.0×, 1.9×, and 1.8× in the

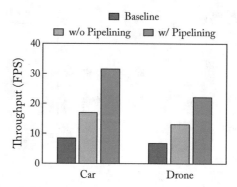

Figure 5.13: FPS of baseline and with and without frontend/backend pipelining.

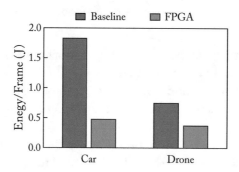

Figure 5.14: Energy per frame comparison between baseline and FPGA design.

three modes, respectively, which leads to a 1.9× overall speedup. The overall standard deviation of latency is reduced by 42.7%. The average throughput is improved from 7.0–22.4 FPS. The average energy per frame is reduced by 47.4% from 0.8–0.4 J per frame. The energy-saving is lower than the Car design because the FPGA static power stands out as the Drone design reduces the dynamic power.

5.6   SUMMARY

In this chapter, we have identified and addressed a key challenge in deploying autonomous machines in the real world: efficient and flexible localization under resource constraints. Using data collected from our commercial deployment and standard benchmarks, it shows that no existing localization algorithm (e.g., constrained optimization-based SLAM [256–258], probabilistic estimation-based VIO [230, 259–261], and registration against the map [256, 262]) is sufficiently flexible to adapt to different real-world operating scenarios. Therefore, accelerating any single algorithm will unlikely be useful in real products. A general-purpose localization algorithm that integrates key primitives in existing algorithms along with its implementation in FPGA is presented. The algorithm retains high accuracy of individual algorithms, simplifies the software stack, and provides a desirable acceleration target.

The unified algorithm framework on FPGAs enables a unified architecture substrate, a key difference compared to prior localization accelerators on ASIC [218, 219, 221] and FPGA [71, 220, 232, 263–266] that target only specific algorithms and scenarios. Additionally, with the same design principle, the architecture can be instantiated to target different autonomous machines with varying performance requirements and constraints, e.g., drones vs. self-driving cars.

CHAPTER 6

Planning on FPGAs

Motion planning is the module that computes how a robot or autonomous vehicle maneuvers itself. The task of motion planning is to generate a trajectory without colliding any obstacles and sends it to the feedback control for physical robot control execution. The planned trajectory is usually specified and represented as a sequence of planned trajectory points. Each of these points contains attributes like location, time, speed, etc. Usually, the motion planning module is followed by the control module, which is used to track the differences between the actual pose and the pose on the pre-defined trajectory by continuous feedback, which has no essential difference with general mechanic control.

Generally, when designing motion planning solutions, one essential requirement is real-time. Motion planning will become a relatively complicated problem when robots work with a high DOF configurations since the search space will be exponentially increased. Typically, state-of-the-art CPU-based approaches take a few seconds to find a collision-free trajectory [267–269], making the existing motion planning algorithms too slow to meet the real-time requirement for complex navigation tasks in complex environments. With massive data-parallel processing, high-performance GPUs can achieve motion plans on the order of hundreds of milliseconds [270, 271] at hundreds of watts in power cost. The performance is still insufficient for real-time requirements, and the power consumption is excessively costly for many robot applications on edge.

Another key design consideration of motion planning solutions is flexibility. Typically, the surrounding in which a robot is situated is complex and dynamic. The position, number, and velocity of obstacles can change with time. Therefore, the constructed roadmap needs to be dynamically updated to reflect the latest status of the environment, which requires the flexibility of practical motion planning and control solutions. However, customized accelerators may not be flexible enough. FPGA-based solutions can be a good compromise between high-performance and flexibility, and have demonstrated some advantages in recent works [272, 273]. In this chapter, we give an overview of robotic motion planning algorithms, then present and summarize their FPGA-based accelerators and discuss the main techniques used.

6.1 MOTION PLANNING CONTEXT OVERVIEW

In robotics and autonomous vehicles, motion planning seeks to find the optimal collision-free trajectory from the start position to the goal position in a dynamic world. It is usually invoked during a robot movement to adapt to environmental changes and continuously search for a safe path to the destination.

Although the motion planning problem is considered to be challenging from a computational point of view [274], several practical algorithms have been proposed in the literature. One of the most widely used classes of practical motion planning algorithms is sampling-based solutions. They provide large amounts of computational savings by avoiding explicit construction of obstacles in the state space and demonstrate empirical success in many scenarios.

Among sampling-based motion planning algorithms, arguably, the most influential solutions to date are Probabilistic RoadMaps (PRMs) [43] and Rapidly exploring Random Trees (RRTs) [42] along with their variants [267]. Both PRM and RRT algorithms provide probabilistic completeness guarantees in the sense that the probability that these algorithms return a solution, if one exists, converges to one as the number of samples approaches infinity [275]. Even though the idea of connecting points sampled randomly from the state space is essential in both approaches, these two algorithms differ in the way that they create a graph connecting these points. Next, we will introduce the workflow of PRM and RRT algorithms. Interested readers are pointed to [267] for more details.

6.1.1 PROBABILISTIC ROADMAP

The PRM motion planning algorithm and its variants are multiple-query methods. Generally, PRM algorithm contains three steps: roadmap construction, collision detection, and graph search.

Roadmap Construction

Roadmap construction is the generation of a graph of poses and motions in the robot's configuration space in an obstacle-free environment. Each vertex in the graph defines a possible configuration state of the robot (e.g., joints' angles). Then these vertices are connected with edges, and each edge defines a possible movement between poses. Common algorithms randomly sample poses from configuration space and build the general-purpose roadmap that is completely independent of obstacles.

Collision Detection

Collision detection checks whether the robot's movement will collide with any obstacle in the environment and whether its components will collide with itself. In a multi-agent robot system, we also need to check if two different robots have the possibility to collide with each other. Triangle meshes are widely used as the models for the surrounding and robot, so collision detection contains checking whether any triangles of the robot model intersect with those of the environmental model, which is quite computationally expensive.

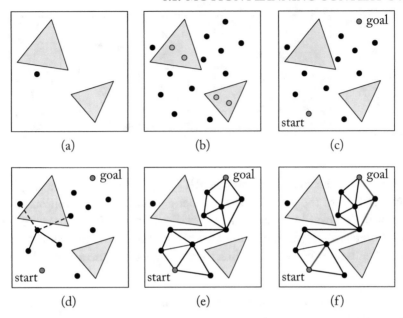

Figure 6.1: Illustrating the main steps of the PRM algorithm. Roadmap construction: (a, b) sample N configurations at random from the configuration space; (c) add all feasible configurations and the start and goal to the roadmap. Collision detection: (d, e) test pairs of nearby milestones for visibility, and add visible edges to the roadmap. Path search: (f) search for a path from the start to the goal.

Path Search

Path search traverses the obtained collision-free roadmap to check whether a path from the start pose to goal exists and determine the optimal path. This process is usually done with Dijkstra's, A*, or Bellman–Ford algorithms and their variants [276].

Figure 6.1 illustrates the main steps of the PRM algorithm. Usually, PRM can reliably solve high-dimensional problems that have good visibility properties, but may suffer from narrow passage problems.

6.1.2   RAPIDLY EXPLORING RANDOM TREE

There exist many types of sampling-based planners that create tree structures of the free-space instead of a roadmap, e.g., RRT. RRT is primarily aimed at single-query applications. It incrementally builds a tree of feasible trajectories, rooted at the start configuration. Figure 6.2 gives the steps of RRT algorithm. The algorithm is initialized with a graph that includes the initial state as its single vertex, and no edges. At each iteration, a point q_{rand} is sampled, and an attempt is made to connect the nearest vertex in the tree to the new sample. If such a connection

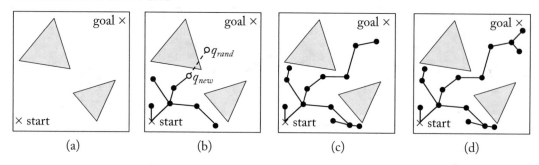

Figure 6.2: Illustrating the main steps of the RRT algorithm. (a) A tree is grown from the start configuration toward the goal. (b) The planner generates a configuration q_{rand}, and grows from the nearest node toward it to create q_{new}. (c) The tree rapidly explores the free space and avoids the detected collision. (d) The planner terminates when a node is close to the goal node. Common implementations will connect directly to the goal. (Figure adapted from [277].)

is successful, q_{rand} is added to the vertex set along with (q_{new}, q_{rand}) added to the edge set. The iteration is stopped as soon as the tree reaches a node in the goal region.

Compared to PRMs, RRTs are the relative ease of integrating motion parameters such as system dynamics, and avoids the necessity to set the number of samples a priori. Common variants of RRT include RRT-Connect that grows two trees of feasible paths, one rooted at the start and the other at the goal [278], or an asymptotically optimal solution to find the shortest trajectory by discarding cycles with a longer path [267]. Other sampling-based motion planning solutions include Expansive Space Trees (EST) [279] and Sampling-based Roadmap of Trees (SRT) [280]. The latter combines the main features of single-query algorithms such as RRT and EST with multiple-query algorithms such as PRM.

6.2   COLLISION DETECTION ON FPGAS

6.2.1   MOTION PLANNING COMPUTE TIME PROFILING

The performance of collision detection is usually the bottleneck of the entire motion planning pipeline. In a practical scenario, motion planning typically involves a large number of collision detections. Take RRT as an example. When it runs on a CPU, 99% of the instructions are executed for collision detection, which takes up over 90% of the total motion planning computation time [281]. More illustrative profiling of the computation time spent in collision checking and graph operation is shown in Fig. 6.3. The motion planning algorithm is executed for one million iterations. The per-iteration time spent on finding the nearest neighbor and inserting a new node, along with the time spent doing collision checking, are demonstrated in Fig. 6.3a. It is clear that collision detection is the bottleneck. The per-iteration time spent on graph operations alone is shown in Fig. 6.3b, illustrating its logarithmic increase. Note that after a million itera-

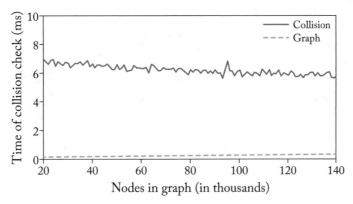

(a) Computation time of collision checking and graph operations.

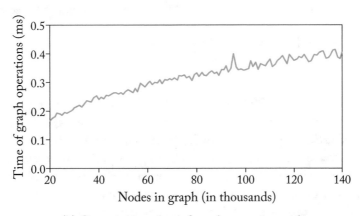

(b) Computation time of graph operations only.

Figure 6.3: Computation time of sub-algorithms. (Figures adapted from [281].)

tions, the RRT sought 40,000 solutions, and yet the graph operations still consumed less than 1% of the time used by the collision checker.

6.2.2   GENERAL PURPOSE PROCESSOR-BASED SOLUTIONS

Several research groups have focused on exploiting the high-performance of GPUs to parallelize and improve the performance of conventional planning algorithms [270, 271, 281]. Bialkowski et al. [281] are concerned with massively parallel implementations of incremental sampling-based algorithm RRT and its asymptotically optimal counterpart RRT*. They break the collision detection tasks of the RRT* algorithm into three parallel dimensions to construct thread block grids that can execute collision computations simultaneously.

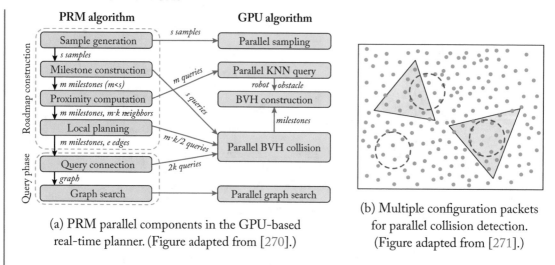

(a) PRM parallel components in the GPU-based
real-time planner. (Figure adapted from [270].)

(b) Multiple configuration packets
for parallel collision detection.
(Figure adapted from [271].)

Figure 6.4: Overview of GPU-based motion planning scheme proposed in [270] and [271].

Pan et al. [270] present a PRM algorithm-based motion-planning framework called *g-Planner*, which is completely implemented on GPUs to avoid the expensive data transfer between the CPU and GPU. They broadly divide PRM algorithm into the roadmap construction phase and query phase, and use a many-core GPU to greatly improve the performance of each component. The framework of the overall GPU-based planner is shown on the right side of Fig. 6.4a. The key idea of this design is to construct efficient algorithms that have linear space and time complexity and exploit the multiple cores and data parallelism effectively.

To further fully utilize the data-parallelism and multi-threaded capabilities of many-core GPU, and its unique programming model and memory hierarchy, Pan et al. [271] present parallel algorithms to accelerate collision queries for sample-based motion planning. This design adopts a clustering scheme and collision-packet traversal to execute efficient collision queries on multiple configurations simultaneously. As shown in Fig. 6.4b, in the configuration space, the green points are random configuration samples and gray areas are obstacles. Configurations adjacent are clustered into configuration packets (red circles) based on their boundary, and configurations in the same packet are then mapped to the same warp on a GPU. In addition, this design introduces a hierarchical traversal technique that can perform workload balancing for high parallel efficiency.

However, although these works have demonstrated that motion planning can achieve substantially higher performance on GPUs than on CPUs with some amount of algorithm tuning, the performance and energy efficiency (performance/watt) are still incompetent for real-time motion planning for complex robots. This problem will be even worse when dealing with robots on edge with limited power supplies. Another limitation of the GPU method is a GPU usu-

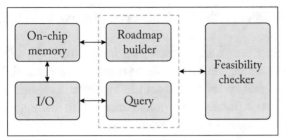
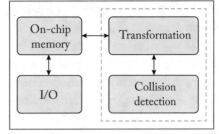

(a) Motion planning configuration, feasibility checker can be specified depending on the problem, e.g., collision detection or potential energy computation.

(b) Dedicated collision detection configuration.

Figure 6.5: Motion planning FPGA-based designs in different configurations from [282].

ally provides only a constant speedup factor; once the GPU cores are completely utilized, the execution time will scale linearly with the size of the problem.

6.2.3 SPECIALIZED HARDWARE ACCELERATOR-BASED SOLUTIONS

The inability to generate paths in real-time of CPU and GPU solutions is a major barrier to the widespread deployment of robotics in the workplace and the home. Much of the recent work has focused on specialized hardware for robot motion planning. Specifically, robot-specific circuitry and architecture are proposed by leveraging algorithm-hardware co-design techniques to fully exploit a combination of aggressive pre-computation and massive parallelism. Generally, specialized hardware designs can generate motion plans for actual robots approximately three orders of magnitude faster than existing GPU implementations [272, 273, 282–284].

FPGA-Based Design 1

Atay and Bayazit focus on directly accelerating the PRM algorithm on FPGAs [282]. The authors present an accelerator design not just for collision detection, but for most of the learning phase of a batch-variant of the PRM algorithm. They create functional units to conduct the random sampling of new configurations, as well as the nearest neighbor search.

The overview diagram of their design is presented in Fig. 6.5. There are five major modules in the proposed FPGA-based motion planning processor (Fig. 6.5a): (i) I/O is responsible for communication between the host computer and the chip; (ii) memory stores the environmental models; (iii) the roadmap builder builds a roadmap using the current models and feasibility criteria; (iv) the query finds a path through the roadmap; and (v) the feasibility checker checks whether the current configuration of the robot satisfies the feasibility constraint. This modular representation can help switch individual components to reconfigure the hardware. For example, collision detection is currently used for feasibility criteria, but the feasibility check module can be switched to potential energy computations in the future. Similarly, by removing roadmap builder and query modules, more space is dedicated to the feasibility of constraint checker and collision detection (Fig. 6.5b).

The parallelism in this design is primarily drawn from implementing multiple triangle-triangle intersection functional units. By conducting multiple pair-wise checks in parallel, it accelerates the checking of whether or not the robot and obstacles collide. However, a significant amount of arithmetic (especially multiplications) is involved in triangle intersection testing, and even high-capacity FPGAs cannot accommodate nearly enough multipliers to make this design feasible. Consequently, in the simplified test case, the authors are unable to fit their design on an FPGA.

FPGA-Based Design 2

Murray et al. [272, 285] present a custom hardware microarchitecture on FPGAs that performs voxel-based collision detection. Assuming that the robot resides in a world without obstacles, they first build a single general-purpose PRM in advance. This roadmap construction scheme differs from conventional sampling-based planners such as PRM or RRT, where these algorithms incrementally build up a roadmap at runtime to navigate through the obstacles present at that time. Then the authors adapt the PRM to a specific instance of problem by removing the edges where collision is detected. Finally, a motion plan is generated by simply finding a path through the resulting smaller PRM.

The detailed process for producing robot-specific motion planning circuitry in [272] are as follows. (a) A robot description is given. (b) A PRM is built where is most likely subsampled for coverage from a much larger PRM. (c) The reachable space of the robot is discretized into depth pixels, and for each edge i on the PRM, all the depth pixels that collide with the corresponding swept volume are pre-computed. (d) These values are used to construct a logical expression that, given the coordinates of a depth pixel encoded in binary, returns true if that depth pixel collides with edge i. (e) This logical expression is optimized and used to build a collision detection circuit (CDC), and there is one such circuit for each edge in the PRM. (f) When the robot needs to create a motion plan, it perceives its environment, determines which depth pixels correspond to obstacles, and transmits their binary representations to every CDC. All CDCs perform collision detection concurrently, in parallel for each depth pixel, storing a bit which indicates that the edge is in collision and should be removed from the PRM. This design can solve a motion planning query within 1 ms, which is adequate for the real-time requirement.

However, there are two main limitations of this implementation. First, the design is constrained by the available hardware resources to represent PRM edges. Relatively small PRMs will consume a substantial portion of the capacity of high-capacity FPGAs. Second, this design is not retargetable to different robots and scenarios. The PRM must be constructed in advance. Consequently, if the robot's physical configuration changes, the planning circuitry may no longer be accurate. Reconfiguring the FPGA takes just a few seconds, which will significantly deduct the real-time performance.

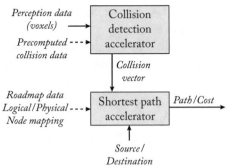

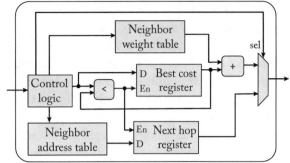

(a) The overall design dataflow. Dotted arrows indicate communication that happens during the programming phase, and solid arrows indicate runtime communication.

(b) Each Bellman Ford Compute Circuit (BFCC) has a table of the physical addresses of its logical neighbors, and a table of their edge weights.

Figure 6.6: The overall dataflow and bellman ford compute circuit module in Murray et al. [273] design. (Figures adapted from [273].)

FPGA-Based Design 3

While fast, the design in [272] is not retargetable to different robots and scenarios. To address this limitation, Murray et al. [273] further propose a programmable architecture for accelerating collision detection and graph search in robotic motion planning.

Similar to the design in [272], a general and large roadmap is first precomputed by leveraging any prior knowledge about the scenario before runtime. The roadmap is made sufficiently large and redundant to be robust to obstacles, allowing successive queries to be completed fast in dynamic environments without reprogramming the accelerator. The collision detection results are also precomputed in a discretized view of the environment during this phase.

There are two main techniques proposed in [273] to make the design flexible and retargetable (Fig. 6.6). First, different from the design in [272, 285] that performs collision detection by building circuits of combinatorial logic which directly correlates to the swept volume of each motion in the roadmap, Murray et al. [273] implement a sea of compute-elements containing registers that are filled with swept-volume data for the robot and roadmap of interest. After configuration, the voxels are sent to the accelerator from the sensed occupancy grid at runtime to be checked against the swept volumes. Then the collision detection results pass to the path search architecture. The overall dataflow of this design is presented in Fig. 6.6a. Second, to handle any expected graph topology, they design a dataflow microarchitecture to perform path search that consists of a sea of circuits connected by a low-cost interconnection network used to handle various topology, referred to as Bellman–Ford Compute Circuits (BFCCs), as shown in Fig. 6.6b.

Every vertex in the graph is statically assigned to a physical BFCC on the chip, and BFCC is small enough that the microarchitecture can scale to large graph sizes.

The programmability enables this architecture to be applied to a wide range of diverse applications for robotics and motion planning. With a modest power consumption of 35 W, this design can perform collision detection and calculate a path in 2.3 microseconds. This latency is approximately five orders of magnitude faster than conventional sampling-based planners, and two orders of magnitude faster with the same use case than [272, 285].

In addition, this design achieves great scalability. The time to execute collision detection is independent of the number of edges in the roadmap since there is dedicated hardware for each edge. The time to perform the shortest path search scales linearly with the number of hops in the graph since there is dedicated hardware for each node.

ASIC-Based Design 1

One major limitation of the aforementioned FPGA-based collision detection accelerators is the relatively few on-chip RAM resources, limiting their capability to store large maps. To solve this issue, Lian et al. [283] propose an accelerator called Dadu-P that stores edge data in memory rather than in circuits, supporting a large roadmap in a dynamic environment and bringing higher flexibility.

The framework of Dadu-P is as follows: first, the algorithm builds up the roadmap in the collision-free environment, where each edge is stored as an octree, representing the swept volume covered by the motion of the robot, which is similar as [272, 273]. Second, the sensor will send every obstacle-covered voxel ID to every edge that will be judged by processing element (PE) of whether it will be affected or not, and the edge will be permitted if not affected. Third, the shortest path search algorithm can be performed to find the solution after a batch of edges has been judged. This process will continue one after another, and can run in single-threaded or dual-threaded.

The architecture of Dadu-P is presented in Fig. 6.7a. It consists of N PEs, where each PE can check the collision for an edge in the roadmap, resulting in N-edge checking in parallel. Every PE contains an on-chip SRAM that stores the 3D model description of the edges in the roadmap. When executing the query, the voxel ID will be sent to the accelerator and broadcasted to all PEs for collision detection. After the detection is completed, the accelerator gathers the results from all PEs and output a result containing N bits, where 1 bit for each PE. To process the next N edges, the accelerator needs to load these edges from off-chip memory to on-chip SRAM.

Dadu-P achieves 360 μs latency for 2500 edge collision detections, increasing the speed by 26.5× and 15.2× compared with the existing CPU and GPU implementations. Note that by storing edge data in memory rather than in circuits, Dadu-P enables reconfigurability, but the external memory transfer indeed causes a 25× latency increase in collision detection compared to design in Murray et al. [272]. The reason is Dadu-P suffers from the memory bandwidth

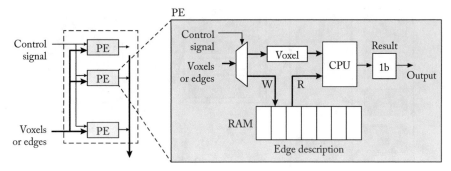

(a) Architecture of the Dadu-P accelerator.

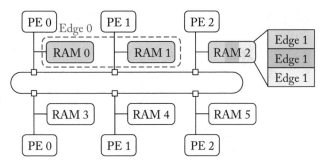

(b) Distributed memory architecture.

Figure 6.7: Overview of architecture and distributed shared memory structure of the Dadu-P implementation. (Figures adapted from [283] and [286].)

bottleneck due to the intrinsic memory-intensive characteristic of collision detection application. For more detailed analysis, Dadu-P has 128 cores running at 1 GHz and requires 37.5 GB of data per second. However, only up to 20 GB of data can be supplied per second due to the limited bus bandwidth of the DRAM, implying that the cores are waiting for data almost half of the time. Such a memory bandwidth bottleneck significantly limits the maximum performance that Dadu-P can be reached. One solution to this memory bottleneck issue is leveraging the distributed shared memory architecture [286]. As shown in Fig. 6.7b, by storing one big edge in multiple RAMs, more number of edges can be stored in the on-chip memory, and the reads from the off-chip memory are reduced, thereby improving the performance.

ASIC-Based Design 2
The memory bandwidth bottleneck significantly impacts the maximum achievable performance of custom accelerators and limits their scalability for higher parallelism. To address this limita-

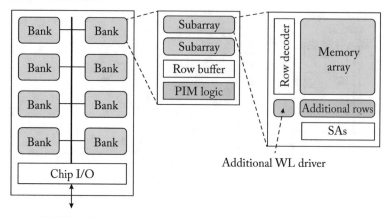

(a) Modifications (colored blue) to the DRAM architecture.

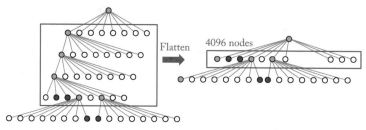

(b) Illustration of a six-level octree and flexible-octree.

Figure 6.8: The memory architecture and logic design that supports processing-in-memory in Dadu-CD implementation. (Figures adapted from [284].)

tion, Yang et al. [284] propose an architecture called Dadu-CD for collision detection by employing the processing-in-memory (PIM) technique, which eliminates the memory bandwidth bottleneck.

The key idea behind PIM is to bring the compute into the memory that can avoid most data movement costs. More background about PIM can be found in [287, 288]. A set of innovative software and hardware strategies are presented in Dadu-CD design to make the algorithm and hardware efficient for PIM, including a PIM logic architecture for massive parallelism, a novel octree data structure for reducing recursions, and a new node encoding method to simplify the PIM hardware design.

Figure 6.8a presents the overall hardware architecture of Dadu-CD. To support PIM operations, a PIM logic module is added to each bank. Additional memory rows and an additional word line (WL) driver are added in each subarray of conventional DRAM architecture. To fully utilize the large bandwidth inside DRAM and avoid bringing too much computing logic, both

Table 6.1: Performance, energy, and design characteristics comparison of different motion planning accelerators

Design	Platform	Time for 2,500 edges	Time for 100,000 edges	Power/ Energy	Area	Scali- bility	Retarget- ability
Murray et al. [272]	FPGA	16 μs	–	15 W	–	low	low
Murray et al. [273]	FPGA	2.3 μs	–	35 W	450 mm²	medium	high
Lian et al. [283]	ASIC	360 μs	14400 μs	472 mW	1.75 mm²	high	high
Yang et al. [284]	ASIC	34 μs	1360 μs	2.13 mJ/check	<2.2% of DRAM	high	high
Murray et al. [272]	GPU	1100 μs	50,000 μs	225 W	–	medium	high
Murray et al. [272]	CPU	9550 μs	380,000 μs	84 W	–	medium	high

edges of the roadmap and obstacles are stored in the same structure. However, conventional octree divides the space recursively which is naturally not suitable for parallelism. To ease massive parallelism, a new data structure called flexible-octrees are presented in Fig. 6.8b, where the top four levels of the octree are flattened into one level while leaving the octree nodes unchanged. Not only does the flexible-octree reduce recursion, but it also improves parallelism.

6.2.4   EVALUATION AND DISCUSSION

Performance Comparison

Table 6.1 summarizes the collision detection latency of various designs presented in this section. Compared to CPU and GPU implementations, [272] achieves 597× and 4152× speed up, respectively. It can meet the real-time requirement for robot operations while with low retargetability. Note that we refer retargetable design here as the ability to deal with various robots and scenarios. [272] create circuits logic directly corresponding to the swept volume of each motion in the roadmap, which brings low programmability and retargetability. FPGA itself can be reconfigured to adapt different scenarios with a few seconds, which will degrade the real-time performance.

[273] implements a programmable architecture and further improve the performance by 7× compared to [272]. [272] is not scalable for the large problem since the FPGA resource is limited to store every edge in the roadmap for dedicated gate-level logic implementation.

Table 6.2: The key design techniques used in motion planning accelerators

Design	Key Design and Optimizing Techniques
Murray et al. [272]	1. Novel hardware-friendly algorithm for massive parallelism 2. Roadmap and collision data pre-compute 3. Robot-specific circuitry and microarchitecture
Murray et al. [273]	1. Roadmap and collision data pre-compute 2. Runtime collision detection data streaming 3. Fully retargetable microarchitecture (Bellman-Ford Compute Circuits)
Lian et al. [283]	1. A hardware-friendly data structure for flexibility 2. Batched processing method to reduce the complexity of algorithm 3. Novel edge sorting strategy to reduce data movement
Yang et al. [284]	1. Processing-in-Memory (PIM) logic design for massive parallelism 2. Novel data structure for promoting parallelism and reducing recursions 3. New algorithm (node encoding scheme) to simplify PIM design

By extensively leveraging off-chip memory to store roadmap in ASIC design, [283] achieves better scalability and higher energy efficiency, while at the cost of 22.5× longer latency. To reduce the expensive data movement, [284] leverages PIM architecture and achieves 10.6× speedup than [283]. In addition, the computation energy and data transfer energy of [284] are significantly reduced by 5.3× and 16× compared to [283]. So there is an intrinsic trade-off between low initial investment and high flexibility of an FPGA, and a high once-off cost but higher memory capacity and power efficiency of ASIC.

Another thing to note that is since there is dedicated hardware for each edge, the time to perform collision detection is independent of the number of edges in the roadmap, which means the custom hardware design is completely parallel and takes constant time to do the computation regardless of the number of edges. By contrast, any given CPU/GPU with a fixed number of hardware threads will experience a linear increase in compute time with the increase of the number of edges.

Design Techniques

These results demonstrate that significant performance improvement is gained by attacking the problem from both sides, i.e., carefully co-design both algorithm and hardware. Table 6.2 summarizes some key design and optimization techniques used in the motion planning accelerators.

From the software aspect, simply mapping existing algorithms to hardware acceleration may not work well, so one solution is to propose hardware-friendly algorithms or data structures to promote massive parallelism and reduce intrinsic recursions, making the algorithm suitable for parallel hardware architecture (e.g., [272, 283, 284]). Another way is to reduce the algorithm

Table 6.3: The time (in μs) taken for path planning in [272] design, broken into five components: identifying goal nodes in the PRM, communication over PCIe, collision detection, deleting edges that are in collision from the PRM, and performing path search. Two representative conventional software approaches (run on a 4-core Intel Xeon CPU) are included for comparison. (Table from [272].)

FPGA-Accelerated Motion Planning						Software (CPU)	
Goal ID	Communi.	Collision	Del. Edges	Path	Total	PRM	RRT
13	118	16	50	425	**622**	815,000	756,000

complexity to achieve higher performance and scalability (e.g., [283]). One more solution is to establish the random roadmap in advance and pre-compute the edges and collision data to reduce the required computation at runtime (e.g., [272, 273, 283]).

From the hardware aspect, one solution is to build robot-specific circuitry and microarchitecture that leverages algorithm characteristics to achieve massive parallel and real-time data streaming (e.g., [272, 273]). Another direction is to reduce data movement, extensive memory access, and complex compute (e.g., [283, 284]). Intelligently mapping algorithms to hardware logic and optimize interconnection networks can also contribute to high-performance and flexible design (e.g., [273]). These techniques can also be leveraged and generalized to other implementations, serving as a guide for future works.

System Analysis

The motion planning process consists of several tasks. Therefore, besides a comparison of collision detection in isolation, we also provide a comparison of the entire motion planning process.

Table 6.3 shows the timing data of FPGA design [272] and conventional software implementations of PRM and RRT. It is well observed that the total time of motion planning (from obstacle data being sent to finishing a motion plan execution) is less than 650 μs on average. And with the several algorithm and hardware optimization techniques, collision detection only takes 16 μs, changing from the bottleneck ($> 90\%$ of total computational time) to one of the fastest components of the whole process. The vast majority of the time is spent on operating systems and the graph search phase, especially Dijkstra's shortest path algorithm, which requires 425 μs and accounts for 68% of the total time. A similar trend is also observed in Dadu-P design, where the Dijkstra graph search algorithm takes up about 92% of the total computation time. This illustrates the removal of the collision detection bottleneck and shifting the distinction to other phases of motion preparation, which motivates algorithm-hardware optimization in graph search that will be briefly discussed in the next section.

6.3   GRAPH SEARCH ON FPGAS

After collision detection, the planner will try to find the shortest and safe path from the start position to the target position based on the obtained collision-free roadmap through graph search. In the previous section, we have demonstrated that the compute bottleneck of the whole motion planning process may shift to graph search after carefully building algorithm and hardware for collision detection. So in this section, we will briefly discuss some FPGA designs for path search.

To find the shortest path during graph search, various algorithms have been proposed. Dijkstra's algorithm [289] and Bellman–Ford algorithm [290] are proposed to solve the single-source shortest-path problem (SSSP). Dijkstra's algorithm is the most efficient sequential implementation in terms of processing speed, but it requires many iteration operations and is hard to parallelize. Bellman–Ford algorithm involves few iterations, and each iteration is highly parallelizable, but each edge can be processed multiple times during the process. In addition to SSSP, Warshall–Floyd Algorithm [291] is proposed to solve the all-pair shortest-path problem (APSP).

Some works leverage general-purpose processors to accelerate the processing speed of the shortest-path problem in terms of optimizing data structure or reducing the computational cost. Harish et al. [294] and Katz et al. [295] implement shortest-path search on the GPU and explore massively parallelized processing, while these designs are not good at serial and complex dataflows. Malewicz et al. [296] use PC clusters with many CPUs to accelerate large-scale graphs, but at the cost of large space and power consumption.

To achieve higher performance and energy efficiency, several studies designed FPGA-based accelerators for graph search by implementing application-specific datapaths. A large amount of work has focused on implementing Dijkstra's algorithm for SSSP on FPGAs [292, 293, 297]. Sridharan et al. [297] propose a parallel algorithm for computing the shortest path between a pair of distinct nodes using Dijkstra's algorithm, along with careful reuse of hardware to achieve time and space efficiency. This design avoids sophisticated data structures and minimizes multiplications and other expensive operations. Takei et al. [292] improve the design for a higher degree of parallelism and large-scale graph search. The memory usage is reduced by a replacement scheme of node data, and the bottleneck of data transfer of the graph data from the external storages to FPGA boards is also reduced. The block diagram of the over architecture is shown in Fig. 6.9. Lei et al. [293] propose a parallel SSSP implementation on FPGA derived from the variant of the Dijkstra algorithm. An extended systolic array priority queue called ExSAPQ is implemented to allow large graph processing (Fig. 6.10). This solution achieves $5\times$ speedup and 1/4 power consumption over the CPU implementation.

Some works also focus on accelerating the Bellman-Ford algorithm for SSSP on FPGAs. As discussed in the last section, Murray et al. [273] accelerate graph search with the Bellman–Ford algorithm. By leveraging a precomputed roadmap and bounding specific robot quantities, this design enables a more compact and efficient storage structure, dataflows, and a low-cost

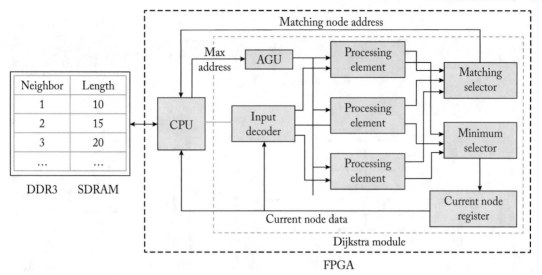

Figure 6.9: The overall architecture of [292] design, consisting of external DDR3 memory, a CPU core, and a Dijkstra module. (Figure adapted from [292].)

interconnection network. Zhou et al. [298] choose to store the entire graph in the DRAM for large graph processing. The key technique in this design is an efficient data forwarding scheme to handle the data hazards in the pipelined architecture, and a data layout scheme to enable efficient utilization of the external memory bandwidth.

Besides SSSP, some researchers accelerate Warshall-Floyd Algorithm for APSP. Bondhugula et al. [299] employ a parallel FPGA-based design using a blocked Floyd–Warshall algorithm [300] to solve large instances of All-Pairs Shortest-Paths (APSP) problem, which achieves a 15× speedup over an optimized CPU-based implementation. The key point in this design is to minimize associated data movement costs on the system.

6.4   SUMMARY

In this chapter, we introduced the motion planning modules of the robotics system, along with several commonly used algorithms. According to the characterization results in Section 6.2.1, collision detection takes up 99% of the CPU instructions and more than 90% of the total motion planning computational time. With careful algorithm-hardware co-design, FPGA can achieve three orders of magnitude than CPU and two orders of magnitude than GPU with much lower power consumption. This demonstrates that FPGA is a promising candidate for accelerating motion planning kernels. We also compare several FPGA and ASIC accelerator designs in motion planning to analyze intrinsic design trade-offs. Several FPGA design techniques have been

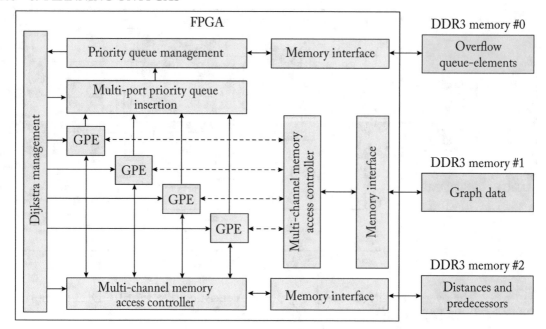

Figure 6.10: The overall architecture of [293] design. The off-chip memory consists of three independent DDR3. The FPGA consists of several modules, including Dijkstra management, graph processing engine, multi-channel memory access controller, multi-port priority queue insertion, and priority queue management. (Figure adapted from [293].)

summarized and discussed, disclosing the benefits of software and hardware co-design and serving as a guide for future work.

There are several directions that need further research efforts. On the one hand, an end-to-end acceleration solution is necessary. Most existing FPGA or ASIC implementations are problem-specific and usually accelerate a specific compute kernel. An end-to-end motion planning and control acceleration solution from system-level that integrates different tasks and target a wide range of robotic applications are necessary for the robotic community. Parallelism and data communication patterns across different tasks need to be investigated. Optimization techniques need to be explored not only from the algorithm compute side, but also from data pre-processing, post-processing, storage, and network bandwidth sides, to achieve a holistic and end-to-end improvement. On the other hand, balancing and optimizing trade-offs always need consideration in the future design, e.g., how should we correctly select the combination of algorithm and hardware, so that the designs can achieve the best balance of performance, energy efficiency, scalability, and flexibility under various scenarios.

C H A P T E R   7

Multi-Robot Collaboration on FPGAs

Thus far, we have focused on the utilization of FPGAs in single-robot applications. In this chapter, we consider collaborative exploration through a team of robots, in which the robots share information with each other or even with the infrastructure [301]. Especially, we discuss how FPGAs can be utilized to accelerate multi-robot acceleration workloads. Through this chapter, the readers shall understand the unique compute challenges in multi-robot exploration workloads and how FPGAs can be utilized to address these challenges.

7.1 MULTI-ROBOT EXPLORATION

Multi-Robot Exploration (MR-Exploration) [302] provides location and map for each robot, and is the basic task for many multi-robot applications, such as multi-robot navigation [303] and multi-robot rescue [304]. An MR-Exploration system [302, 305] consists of several robots, and each robot executes the system illustrated in Fig. 7.1. Each input frame is fed to the Feature-point Extraction (FE, ①) module for Visual Odometry (VO). FE and VO are also the basic components of single-robot applications.

Some input frames, called keyframes, are fed to the Place Recognition (PR, ②) module. The PR module generates the compact image representation, which produces the candidate place recognition matches between different robots. PR connects different robots through the same place, and it is a main added module of multi-robot systems than single-robot systems. The Visual Odometry (VO) (③) matches the feature-points of two adjacent frames to produce the 6-DoF poses. Although the VO outputs the 6-DoF poses, which is represents the pose in 3D space, we build the 2D map only using several lines of the camera. By projecting the 3D poses into 2D maps to navigate the robots, we significantly reduce the computational cost of building maps. Furthermore, each 2D map is aligned to a keyframe [306, 307]. Through optimizing the keyframe trajectory poses, the map can be synchronously optimized and merged. The relative pose (RelPose) module does the same operation as VO(③) and establishes relative poses between the candidate place matches of different robots.

The decentralized optimization module (DOpt, ④) optimizes the intra-robot relative pose measurements from VO and the inter-robot relative pose measurements from RelPose. In this chapter, we show how we use the Pose Graph Optimization (PGO) [308] to do DOpt. PGO only optimizes the keyframe trajectory pose without optimizing the keypoints. Because

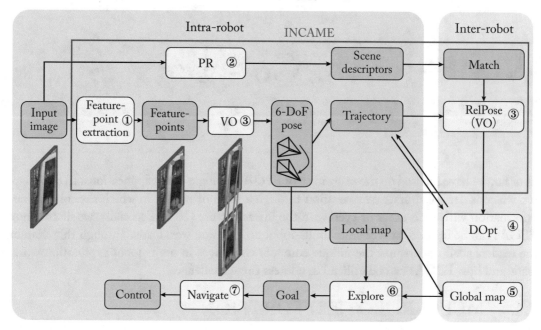

Figure 7.1: The components in MR-Exploration. Each frame is required to execute ①③ with low latency. ② runs on some keyframes. INCAME implements ①② on the CNN accelerator, and optimizes the scheduling among ① ② ③.

the number of keyframes is far less than the number of feature points, PGO has been demonstrated much faster than Bundle Adjustment (BA) algorithm [309], which optimize not only the trajectory but also the feature points. The global map generation module (⑤) merges the maps. The exploration module (⑥) decides an unexplored goal point for each robot to move based on the merged map and the estimated location. The navigation module (⑦) guides each robot to the goal point, including path planning and obstacle avoidance.

With the help of 2D maps (⑤) and fast PGO (④), we deploy map merging and optimization on the embedded CPU. PR (②), feature-point extraction (①), and VO (③) become the most computationally intensive modules. We optimize the scheduling of these components across the CPU side (processing system, PS) and FPGA side (programmable logic, PL) on Xilinx MPSoC [310].

Recent works use CNN to extract feature points [311–313] and generate the place representation code [72, 314]. Compared with the popular handcrafted extraction method, ORB [315], the CNN-based feature point extraction method, SuperPoint [311], achieves 10–30% higher matching accuracy. The accuracy of the place recognition code based on another CNN-based method, GeM [72], is also about 20% better than the handcrafted method, root-SIFT [73].

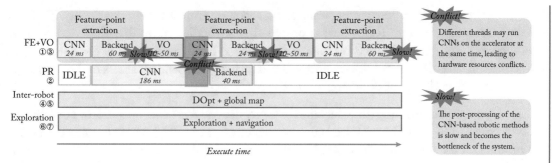

Figure 7.2: The overall timeline of MR-Exploration. The FPGA accelerator is adapted in the feature-point extraction (FE) and place recognition (PR). The simultaneous occupancy of a single FPGA accelerator by these two software threads will lead to hardware resource conflicts. The post-processing of the visual odometry (VO) with feature-points on embedded CPU is time-costing and cannot meet the real-time requirements.

For these reasons, CNN is increasingly being deployed in robotic systems. Besides these two components, more CNN-based methods, such as semantic segmentation [316] and object detection [317], can be introduced into robots to achieve better performance in the future. On the other hand, CNN is computationally expensive. A single inference of the CNN-based SuperPoint feature-point extraction consumes 39 G operations [311], while a single inference of the CNN-based GeM place recognition consumes 192 G operations [72]. Therefore, specific hardware architectures on FPGA [90, 92, 318–320] are designed to deploy CNN on the embedded system. With the help of neural network quantization and on-chip data reuse, the speed of CNN accelerators on embedded FPGAs achieves 3TOP/s [320], which can support the real-time execution of CNN-based feature-point extraction [311]. However, these FPGA-based CNN accelerators are designed and optimized to accelerate a single CNN. They cannot automatically schedule two or more tasks simultaneously.

In this chapter, we present a simple profiling of MR-Exploration with the FPGA-based CNN accelerator. The CNN backbones of FE and PR are executed on FPGA accelerators (Angel-Eye [92]). Other operations, including the post-processing of the CNN-based FE and PR, run on the PS side of Xilinx ZCU102 evaluation board [321]. The timeline of each component is illustrated in Fig. 7.2. The threads of FE and PR may need to process CNN at the same time, and the simultaneous requests of the accelerator will lead to hardware resource conflicts. For FPGA-based CNN accelerators, the inability of multi-task makes it difficult for researchers in robotics to use embedded FPGA.

In order to facilitate robotic researchers to run different CNN tasks simultaneously on FPGA accelerators, an FPGA accelerator should support the following features.

Multi-Thread: Different components in a robot are from different developers. Thus, the Robot Operating System (ROS) [63] is proposed as a middleware to fuse these independent components, and is widely used by robotic researchers. Each component is considered as an independent thread in ROS. Different threads should have independent access to the accelerator without knowing the status of others.

Dynamic Scheduling: The execution of CNN depends on other operations. For example, CNN-based FE (① in Fig. 7.1) needs to be executed after the VO (③) is completed. The VO is running on the CPU, and the execution time varies with the input data [322] (10–50 ms in Fig. 7.2). The accelerator cannot predict when to start a task. Therefore, the FPGA accelerator should be scheduled dynamically to support irregular task requests from the software.

Scheduling by Priority: Different components have different priorities. The control and perception tasks usually have higher priorities, while the long-term decision and optimization have lower priorities [323]. The critical tasks, which are latency-sensitive, need to be processed firstly on the accelerator. The concept of interrupt [324] was introduced to CPU in the 1960s. It enables the CPU to support dynamic multi-task scheduling according to priority to satisfy these three functions. Therefore, we introduce the concept of interrupt to the CNN accelerator in this chapter to support multi-task on FPGA-based accelerators.

Besides the CNN backbones, the post-processing of the CNN-based methods, including normalization, softmax, rank, etc., are also computationally intensive. As illustrated in Fig. 7.2, the execution time of post-processing for FE on embedded CPU (~60 ms) exceeds that of CNN backbone on the accelerator (~30 ms), and becomes the bottleneck of the system. As mentioned before, CNN is widely used in a variety of robot tasks. The post-processing of these different tasks may become the new bottleneck of the system. It is necessary to use FPGA to speed up post-processing, and a framework is also needed to organically integrate the CNN backbones and post-processing operations.

To address the above challenges, we introduce an INterruptible CNN Accelerator for Multi-robot Exploration (INCAME) for rapid deployment of robot application on FPGAs. The details are listed below.

- We present a CNN-based MR-Exploration framework, INCAME. CNN-based methods for feature-point extraction (FE) and place recognition (PR) are accelerated with FPGA on the ROS platform [63]. With the help of the unified interface in ROS, these CNN-based methods can be easily used by other developers in different applications.

- We present a **virtual-instruction-based** interrupt method to make the CNN accelerator support dynamic multi-task scheduling by priority. The method solves the hardware resource conflicts when accelerating different CNN tasks on ROS.

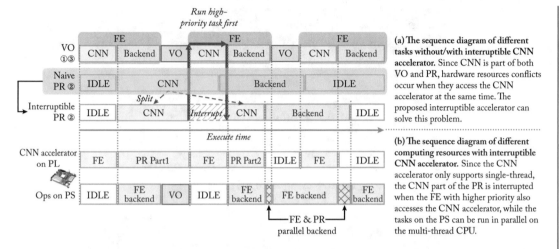

Figure 7.3: Illustration of the accelerator interrupt to solve the hardware resource conflicts. When a high-priority task (FE) starts before the low-priority task (PR) completes, the CNN accelerator backs up the status of the low-priority task to the memory, and processes the high-priority task. When the high-priority task finishes, the low-priority task resumes and continues.

- We present the data flow of the post-processing operations. Hardware modules are also designed for optimized post-processing. This software-hardware co-optimization assures the real-time performance of MR-Exploration.

The rest of this chapter is organized as follows. Section 7.2 introduces the INCAME framework with ROS. Section 7.3 details the virtual-instruction-based interrupt. Section 7.4 gives the evaluation of the interrupt method and gives an example of multi-robot exploration on FPGA. Section 7.5 concludes this chapter.

7.2   INCAME FRAMEWORK FOR MULTI-TASK ON FPGAS

To solve the hardware resource conflicts when the ROS software accesses the CNN accelerator, we propose the INCAME framework in this section. The basic idea to deal with hardware resource conflicts is enabling interrupts on the CNN accelerator, as illustrated in Fig. 7.3.

7.2.1   HARDWARE RESOURCE CONFLICTS IN ROS

Introduction to ROS

Building a real robot requires many different components, including sensors, perception algorithms, and control units from different developers. The Robot Operating System (ROS) [63] is proposed to fuse the components from different researchers into a real system.

```
1. def FEcallback(inputframe):
2.     FECnn = CnnAcc.run(inputframe)
3.     result = FEpostprocess(FECnn)

4. def main():
5.     CnnAcc= InitAccelerator(priority=0)
6.     Subscriber = Node.subscribe(InputframeTopic, FEcallback)
7.     Subscriber.Spin()
```

(a) ROS example for FE node.

```
1. def PRcallback(inputframe):
2.     ... // run CNN and post-processing for PR

3. def main():
4.     CnnAcc= InitAccelerator(priority=1)
5.     ... // Subscribe to topic
```

(b) ROS example for PR node.

Figure 7.4: ROS code examples. Each software function is packaged as an independent *Node* in ROS. A node subscribes to the topics which contain the data structures and processes these data with callback functions (*FEcallback* in (a), (*PRcallback* in (b)). Different nodes do not know the running status of others. In INCAME, the developer only needs to provide the priority (*priority* = 0 or 1) of each task and the hardware schedules these tasks. The FPGA accelerator for the post-processing of FE uses the ScratchPad memory to directly use the CNN results (*FECnn* in (a)) without memory copy.

Each function module, such as FE, PR, and VO, is called a **Node** in ROS. Each node is an independent thread running on the CPU and does not know the running status of others.

A node can publish **ROS topics** and subscribe to topics. The publishing and subscribing nodes connect to the same topic to transmit messages. The subscribing node processes the received topics with callback functions. *Line 6 and 7* in Fig. 7.4a bind the topics (*InputFrame*) with the callback function (*FEcallback* to extract the feature-points). In the callback function, the ROS node processes the CNN backbone to generate the intermediate data (*FECnn*), and executes the post-processing whose input is the generated intermediate data. If a new topic arrives, but the callback function has not finished processing the previous topic, the new topic will be cached in the message queue of ROS.

Hardware Resource Conflicts In ROS

Although ROS is becoming the fundamental software platform for robotics, the independence between different ROS nodes brings **hardware resource conflicts challenge** to access the hardware accelerator. Figure 7.4a *Line 5* and Fig. 7.4b *Line 4* initialize the accelerator for the nodes. *Line 2* in Fig. 7.4a,b runs the CNN backbone on the accelerator, respectively. Different nodes in ROS initialize and run the CNN accelerator independently, which may result in hardware resource conflicts. To address this problem, we set the priorities of different tasks at the initialization phase (the priority parameter), and enable the accelerator interrupt to schedule the high-priority task first.

Accelerator Interrupt

Figure 7.3 illustrates the idea of interrupt to schedule two CNN tasks. In the process of running a low-priority network (PR), the software may send an execution request for the high-priority task (FE). The interrupt enables the CNN accelerator to back up the running state of the low-priority PR network. Then the accelerator switches to the high-priority FE network. After the high-priority task (FE) completes, the low-priority task (PR) is restored to the accelerator and continues to execute.

7.2.2   INTERRUPTIBLE ACCELERATOR WITH ROS (INCAME)

Figure 7.5a illustrates the proposed two-step INCAME framework for mapping ROS-based software to embedded FPGA. The first step is task decomposition, which decomposes the computation in ROS nodes into different INCAME computation types, including CNN backbones, CNN post-processing, and other CPU tasks.

The second step is to deploy the computing onto FPGAs. The CNN backbones of different tasks, such as the VGG model in SuperPoint feature point extraction [311] and the ResNet101 model in GeM place recognition [72], are compiled to the interruptible Virtual-Instruction Instruction Set Architecture (VI-ISA), which runs on the CNN accelerator. The VI-ISA is a simple extension of the original ISA, in which the extension method is not limited to a specific original ISA. Thus, the virtual-instruction-based interrupt can be easily applied to various instruction-based CNN accelerators [90, 318], such as Angel-Eye [92] and DPU [325]. There are some new *virtual* instructions in the VI-ISA, which is responsible for backup and restore the CNN running status and data. These virtual instructions are only executed when an interrupt occurs, otherwise, they are skipped and discarded.

Hardware modules are implemented for the CPU-intensive Softmax [326] and Normalization [327]. Some task-related software optimizations, such as Ranking and Non-Maximum Suppression (NMS) [328], as well as other ROS tasks written in C++/Python, are processed on the CPU side. To eliminate memory copies between CPU cores and CNN accelerators, we use low-latency ScratchPad memory [329] to directly feed the results from CNN backbones to the post-processing modules.

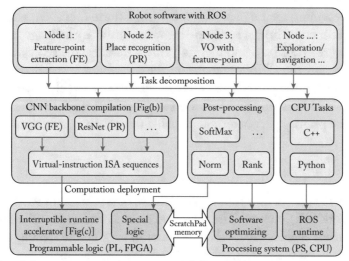

(a) At **task decomposition** step, the operations in different ROS nodes are separated to CNN backbones, CNN post-processing, and other CPU tasks. At **computation deployment**, the CNN backbone and post-processing are deployed with hardware modules and software optimizing.

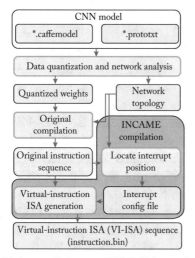

(b) At **compilation step**, INCAME locates the interrupt location, adds virtual instructions, and generates the virtual-instruction ISA (VI-ISA) sequence.

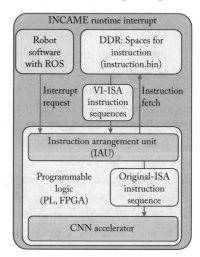

(c) At **runtime**, IAU fetches VI-ISA instructions from DDR and translates the VI-ISA sequence to original ISA sequence executed on the accelerator.

Figure 7.5: Figure (a), INCAME framework. CNN backbones are arranged onto the CNN accelerator. Post-processing operations are arranged onto the special logic. For CNN backbone, INCAME inserts virtual instructions at compilation stage (Fig. (b)), and dynamically schedules different tasks at runtime (Fig. (c)).

Figure 7.5b details the INCAME compilation stage. At the compilation stage, INCAME translates the CNN model from a software framework to the instruction sequence, which can be executed in the interruptible CNN accelerator. Caffe [81] is a popular software framework for CNN, and the *.caffemodel/*.prototxt files define the network parameters and structure in Caffe. The previous deployment processes, such as Angel-Eye [92] and DPU [325], quantize the weights and analyze the network topology. The original compiler translates the network topology and the quantization information into the original ISA sequence. INCAME goes further compared to the previous CNN compilers. It selects the optimized interrupt positions in the original instruction sequence, and adds virtual instructions at these positions to enable accelerator interrupt. After that, the original instruction sequence and the added virtual instructions are wrapped to the new interruptible VI-ISA. The wrapped VI-ISA instructions are dumped into a file (instruction.bin) and can be loaded into the instruction spaces on FPGAs' DDR.

As illustrated in Fig. 7.5c, at runtime, Instruction Arrangement Unit (IAU), a piece of hardware on the FPGA fabric together with the CNN accelerator, listens to the interrupt request from ROS software, fetches the corresponding VI-ISA interruptible instructions and translates them to the original ISA executed on the CNN accelerator in Angel-Eye [92]. The detail of the Virtual-Instruction ISA (VI-ISA) and instruction arrangement unit (IAU) is introduced in Fig. 7.3. Although INCAME can be applied to various instruction-based CNN accelerators, we implement and evaluate it based on Angel-Eye [92].

7.3   VIRTUAL INSTRUCTION-BASED ACCELERATOR INTERRUPT

Chapter 3 shows that CNN is one of the core components for robots and introduces the FPGA-based accelerators for CNN. Based on the previously introduced FPGA-based CNN accelerators, this chapter introduces the support of multi-tasking on FPGA-based CNN accelerators. The multi-robot exploration task can be implemented with the help of the multi-tasking FPGA-based CNN accelerators.

7.3.1   INSTRUCTION DRIVEN ACCELERATOR

As introduced in Chapter 3, there are three categories of instruction in the instruction-driven accelerator: LOAD, CALC, and SAVE [90, 92, 318]. The instruction description of each kind of instruction is listed in Fig. 7.6a and Table 7.1.

The LOAD instruction moves input feature maps and weights from DDR to on-chip memory. The SAVE instruction moves the calculated output features from on-chip memory to DDR.

Each CALC instruction, including CALC_I and CALC_F, processes the convolution according to the hardware parallelism with P_h lines from P_{in} input channels to P_{out} output chan-

Table 7.1: Description for the basic instructions

Category	Type	Description	Address 1	Address 2	Address 3	Workload	Backup	Recovery
LOAD	LOAD_W	Load weights/bias from DDR to on chip weight buffer	Off-chip addr	Weights-buffer addr	–	Data length	–	Weight/input data
	LOAD_D	Load input data from DDR to on-chip data buffer	Off-chip addr	Data-buffer addr	–	Data length	–	Weight/input data
CALC	CALC_I	Calculate intermediate results (from partial input channels) for some output channels from partial input channels.	Input data addr	Intermediate data addr	Weight addr	Calc size	Previous final results/intermediate data	Weight/input data/intermediate data
	CALC_F	Calculate the results for some output channels from all input channels. The pooling, bias-adding, and element-wise operations are operated in this instructions.	Input data addr	Output data addr	Weight addr	Calc size	Final results	Input data
SAVE	SAVE	Save the results from on-chip data buffer to DDR.	Off-chip addr	Data-buffer addr	–	Data length	–	Input data

Type	Address 1	Address 2	Address 3	Workload

(a) The original instruction set for CNN accelerator.

					2bit	16bit
Type	Address 1	Address 2	Address 3	Workload	Virtual	SaveID

(b) The extended instruction set for virtual instruction interrupt method.

Figure 7.6: Figure (a), Original ISA. Figure (b), Virtual-Instruction ISA (VI-ISA).

nels. P_h, P_{in}, and P_{out} are the parallelism along the height, input channel, and output channel dimensions, which is determined by the hardware and original ISA.

Figure 7.7a illustrates the operation of CALC instructions. The convolution of the last P_{in} input channels is CALC_F, and the convolutions for the former input channels are CALC_I. The CALC_F and the CALC_I instructions for the same output channels, as well as the LOAD instructions for corresponding input feature maps and weights, are considered as a **CalcBlob** (Section 7.3.7 lists an example for CalcBlob). In each CalcBlob, there is a LOAD_W instruction for the corresponding weights. However, when there are only few input channels (such $Ch_{in} = P_{in}$), all input data required by a CalcBlob can be fetched to the chip by a single LOAD_D. During the CalcBlob execution, the input buffer remains unchanged. For the next CalcBlob, since the required input data is exactly the same, there is no need to execute a LOAD_D. Thus, some CalcBlobs do not have LOAD_D instruction.

7.3.2   HOW TO INTERRUPT: VIRTUAL INSTRUCTION

As illustrated in Fig. 7.7e, there are four stages to handle interrupt, including: (1) time for finishing the current operation, $t1$; (2) time to backup, $t2$; (3) time for the high-priority task, $t3$; and (4) time to restore the low-priority task, $t4$; The latency to respond the interrupt is:

$$t_{latency} = t_1 + t_2. \tag{7.1}$$

The extra cost for interrupt is:

$$t_{cost} = t_2 + t_4. \tag{7.2}$$

For the instruction flow illustrated in Fig. 7.7c, the interrupt stages are shown in Fig. 7.7e. There are different methods to implement interrupt in FPGA-based CNN accelerators.

CPU-Like. When an interrupt request occurs in CPU, CPU backs up all the on-chip registers to DDR. However, there are only tens of registers in CPU, and the volume of the backed-up data is less than 1 KB [330]. In FPGA-based CNN accelerators, there are hundreds of KB ∼ several MB on-chip caches [90, 92] to store input feature maps or weights. Thus, the extra data transfer increases both the interrupt response latency ($t_{latency}$) and the additional cost (t_{cost}).

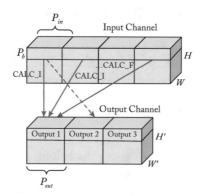

(a) Scheduling of instruction-driven CNN accelerator. The final results of Para$_{out}$ output channels are calculated (solid arrows) before the next output channels (dashed arrow).

(b) Single Blob Save. One SAVE instruction for each CalcBlob's output.

(c) Multiple Blob Save. One SAVE instruction for two or more CalcBlobs' output.

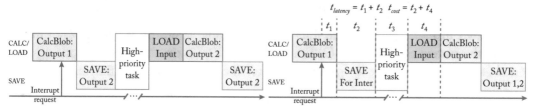

(d) Interrupt for Single Blob Save. The High-priority task starts after the SAVE (*SAVE Output1*) instruction.

(e) Interrupt for Multiple Blob Save. A virtual SAVE becomes valid (*SAVE For Inter*) to backup/store the finished CalcBlob's output.

Figure 7.7: Illustration of scheduling on the CNN accelerator. Figure (a), a CalcBlob is the instructions that calculate the final results of P_{out} output channels from all input channels (the set of solid arrows). Figure (b), one SAVE instruction is only responsible for saving the results of one CalcBlob. Figure (c), one SAVE instruction is responsible for saving the results of several CalcBlobs. Figures (d) and (e) illustrate the accelerator interrupt of Figs. (b) and (c). The latency ($t_{latency}$) and extra cost (t_{cost}) are labeled in Fig. (e).

Layer-by-Layer. Most accelerators run the CNN layer by layer [90, 92]. There is no extra data transfer for the accelerator to switch between different tasks after each layer, thus, $t_{cost} = 0$. However, the position of the interrupt request is irregular and unpredictable. When an interrupt occurs inside a CNN layer, the CNN accelerator needs to finish the whole layer before switching, which leads to the high response latency ($t_{latency}$).

We propose the **virtual-instruction-based** method (*VI* method) to enable low-latency interrupt. Our virtual-instruction-based method is interruptible inside each layer, which can reduce the interrupt response latency. We add some virtual instructions to the original instruction sequence to enable the interrupt. The virtual instructions, which contain the backup and recovery instructions, are responsible for backing up and restoring on-chip caches.

Virtual SAVE instructions back up the intermediate results from partial input channels or the final output results. There is no need to back up the input feature maps and weights because these inputs are already stored in DDR.

Virtual LOAD instructions restore the input feature maps from DDR to on-chip caches.
Virtual LOAD instructions also need to restore the intermediate results from partial input channels backed up by the virtual SAVE instructions.

7.3.3   WHERE TO INTERRUPT: AFTER SAVE/CALC_F

The virtual instruction-based method has two potential factors that may lead to system performance degradation. (1) The extra data transfer to backup/restore running status takes up additional bandwidth resources. (2) The instruction fetching for the virtual instructions also uses bandwidth resources. To address the above problems of virtual-instruction-based method, we analyze the interrupt cost and select the positions of adding the virtual instructions. The backup/recovery data for different interrupt positions at each kind of instruction are listed in the Backup/Recovery columns of Table 7.1. The backup/recovery data transfer for each instruction is analyzed as follows.

LOAD_W/LOAD_D. When an interruption occurs at LOAD, the newly loaded data are immediately flushed when running the high-level CNN, leading to bandwidth waste.

CALC_I. When an interrupt occurs at CALC_I, the unsaved final results (generated by previous CALC_F) should be saved to DDR. The intermediate data from current CALC_I should also be sent to DDR for further use. At the Recovery stage, the intermediate data should be fetched from DDR. The data movement of intermediate results leads to additional bandwidth requirements.

CALC_F. When an interrupt occurs at CALC_F, there are no intermediate results. Although it is necessary to back up the unsaved final results which are generated by previous CALC_F, these results will be stored in DDR through the subsequent original SAVE instruction. If the accelerator can record the interrupt status, we can modify the address and workload when executing subsequent original not-virtual save instruction. In this way, we can avoid the repetitive transmission of the final output results. The input data are shared across the CalcBlobs. Thus, the recovery virtual instruction needs to restore the shared input feature maps.

SAVE. The overhead of interrupt is only to transfer input data from DDR to the on-chip caches.

In order to minimize the cost of interrupt, we make the CNN interruptible after the SAVE or CALC_F. This method only introduces extra data transfer to recovery input data without any extra backup data ($t_2 = 0$). Thus, $t_{cost} = t_4$, in our virtual-instruction-based interrupt.

7.3.4   LATENCY ANALYSIS

In this subsection, we analyze the impact of interruptible position on the interrupt respond latency.

As introduced in Section 7.3.1, each CALC instruction processes the convolution according to the hardware parallelism with P_h lines from P_{in} input channels to P_{out} output channels. The computation time of each pulse is related to the hardware architecture and the width of the convolution layer. The larger the width, the larger the workload of a single calculation instruction, and thus the CALC instruction consumes more time. We note the computation of a CALC instruction a *pulse* and the time consumption of a pulse as a function of the feature map width, t_{pulse}:

$$t_{pulse}(W) = t_{hw} \times W \tag{7.3}$$

t_{hw} indicates the time for hardware to produce one pixel in the output results, which is defined by the architecture and the clock frequency. W is the width of the featuremaps, indicating the workload of the instruction.

The worst interrupt response latency in the Layer-by-Layer method is to wait from the beginning of a layer until the whole layer finishes. The calculation of the whole layer consists of N_{pulse}^{lbl} successive CALC instructions, which is related to the workload (Ch_{in}, Ch_{out}, H) and the hardware parallelism (P_{in}, P_{out}, P_h)

$$N_{pulse}^{lbl} = \frac{Ch_{in} \times Ch_{out} \times H}{P_{in} \times P_{out} \times P_h}. \tag{7.4}$$

We note the worst time of waiting to finish the current layer, which is the total time of these pulses, as t_1^{lbl}:

$$t_1^{lbl} = N_{pulse}^{lbl} \times t_{pulse}(W). \tag{7.5}$$

Ch_{in} and Ch_{out} is the number of input channels and output channels. H is the height of featuremaps.

On the other hand, if the execution of the CNN accelerator can be interrupted after SAVE/CALC_F instructions, the worst case of waiting for finishing the current operation is

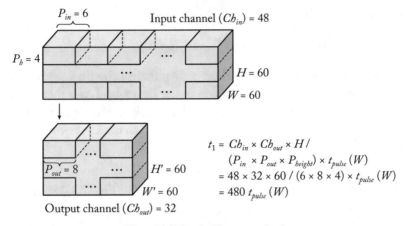

(a) t_1 for layer-by-layer method.

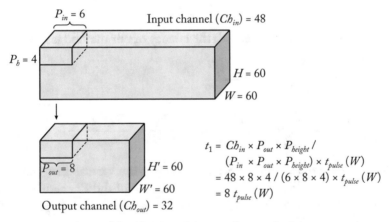

(b) t_1 for virtual-instruction method.

Figure 7.8: An example of waiting time for finishing the current operation (t_1) in a convolution layer. Compared with the Layer-by-Layer method, the waiting time of our Virtual-Instruction method is reduced to 1.6% in this example. The reduction in latency is related to the height (H) of the input feature maps.

illustrated in Fig. 7.8b. The calculation of the whole CalcBlob consists of N_{pulse}^{VI} successive CALC instructions in our Virtual-Instruction method (*VI* method):

$$N_{pulse}^{VI} = \frac{Ch_{in} \times P_{out} \times P_h}{P_{in} \times P_{out} \times P_h}. \tag{7.6}$$

We note the worst waiting time of our VI method as t_1^{VI}:

$$t_1^{VI} = N_{pulse}^{VI} \times t_{pulse}(W). \tag{7.7}$$

The backup operation in the *VI* method only transfers the final results, which are also transferred to DDR with the SAVE instructions in the Layer-by-Layer method. Experimental results, which will be given in Section 7.4.2, show that the data transfer time for the final results is much less than the calculation time (less than 20%), in both the Layer-by-Layer method (t_2^{lbl}) and the *VI* method (t_2^{VI}):

$$t_2^{lbl} \ll t_1^{lbl} ; t_2^{VI} \ll t_1^{VI}. \tag{7.8}$$

Thus, the latency to respond to the interrupt request ($t_{latency}$ in Eq. (7.1)) is mainly determined by the time of finishing the current operation (t_1). As the interrupt request is unpredictable, we model the interrupt location as evenly distributed within each layer. Thus, the average interrupt latency is $\bar{t}_{latency}$:

$$\bar{t}_{latency} \simeq \frac{1}{2} \times t_1. \tag{7.9}$$

Compared with the Layer-by-Layer method, the latency of our method is reduced to R_l:

$$R_l = \frac{\bar{t}_{latency}^{VI}}{\bar{t}_{latency}^{layer}} \simeq \frac{\frac{1}{2} \times t_1^{VI}}{\frac{1}{2} \times t_1^{layer}}$$

$$= \frac{N_{pulse}^{VI}}{N_{pulse}^{lbl}} = \frac{Ch_{in} \times P_{out} \times P_h}{Ch_{in} \times Ch_{out} \times H} = \frac{P_{out} \times P_h}{Ch_{out} \times H}. \tag{7.10}$$

$\bar{t}_{latency\_VI}$ and $\bar{t}_{latency\_VI}$ are the average interrupt latency of the Virtual-Instruction method and the Layer-by-Layer method. The effect of latency reduction of the *VI* method is related to the number of output channels (Ch_{out}) and featuremap height (H). The larger the featuremaps output channels and the height, the better latency reduction result can be achieved.

An example of a convolution layer with a typical size in CNN is given in Fig. 7.8. The parameters are labeled in the figures ($P_{out} = 8, P_h = 4, Ch_{out} = 32, H = 60$). According to Eq. (7.10), the latency can be reduced to $R_t = (P_{out} \times P_h)/(Ch_{out} \times H) = (8 * 4)/(32 * 60) = 1.6\%$. More results will be given in Section 7.4.2.

7.3.5   VIRTUAL INSTRUCTION ISA (VI-ISA)

We add two fields to the instruction set: (1) Virtual and (2) SaveID, as illustrated in Fig. 7.6b.

Virtual Field. The virtual instructions should be only valid when the interrupt occurs. So we add a field in the original ISA, that indicates whether the instruction is a virtual instruction. If no interrupt occurs, virtual instructions will be skipped and discarded, which can ensure the efficiency of uninterrupted execution. Three values can be set to Virtual Field.

2'b00 indicates this instruction is not virtual, should always be executed.

2'b01 indicates this instruction is the SAVE instruction for backup. When an interrupt occurs, the high-priority network will start after this instruction.

2'b10 indicates this instruction is the LOAD instruction for recovery. The corresponding instructions will be executed after the high-priority network.

SaveID Field. SaveID links CalcBlob instructions to the corresponding SAVE. SaveID of each not-virtual SAVE instruction differs. If the generated outputs of CalcBlobs are stored to DDR by a SAVE instruction, the CalcBlobs have the same SaveID as the SAVE instruction.

One SAVE instruction can correspond to one CalcBlob (Single Blob Save, illustrated in Fig. 7.7b) or multiple CalcBlobs (Multiple Blob Save, illustrated in Fig. 7.7c).

For Single Blob Save, no virtual SAVE is added. The high-priority network can be started after the original not-virtual SAVE. The virtual LOAD instructions for data recovery are generated after the original SAVE, and executed after the high-priority network. The execution timeline is shown in Fig. 7.7d.

For Multiple Blob Save, virtual SAVE and LOAD instructions are generated after the CALC_F of each CalcBlob. When the interrupt request occurs, the virtual SAVE instruction will be executed before the start of the high-priority network. Virtual LOAD instructions for data recovery are executed after the high-priority network. The subsequent original not-virtual SAVE instruction with the same SaveID as the CalcBlob will be modified to avoid duplicate output data transfer. The execution timeline is shown in Fig. 7.7e.

7.3.6   INSTRUCTION ARRANGEMENT UNIT (IAU)

Instruction Arrangement Unit (IAU) is the hardware to handle the computing requirements of different priority tasks. The IAU monitors the interrupt status and generates the original ISA instruction sequence. The original CNN accelerator does not need to know the interrupt status and only operates the instructions provided by IAU.

The hardware implementation of IAU is shown in Fig. 7.9, which supports four tasks with different priorities. Task 0 has the highest priority and is not interruptible. InstrAddr records the address to fetch the instructions of the corresponding task. The InputOffset and the OutputOffset, which indicate base address offsets of the input and output data, are configured by the software. SaveID, SaveAddr, and SaveLength record the status when an interrupt occurs. Subsequent not-virtual SAVE instructions will be modified according to the recorded interrupt status (SaveID, SaveAddr, and SaveLength), to avoid duplicate output data transfer.

7.3.7   EXAMPLE OF VIRTUAL INSTRUCTION

The example is based on a straightforward convolutional layer, which has only one input channel and two output channels. The convolution kernel size is 1×1. The shape of the input and output featuremaps is 2×16 (Fig. 7.10a). The parallelism of the CALC instruction in this example is $P_{in} = 1$, $P_{out} = 1$, and $P_h = 2$.

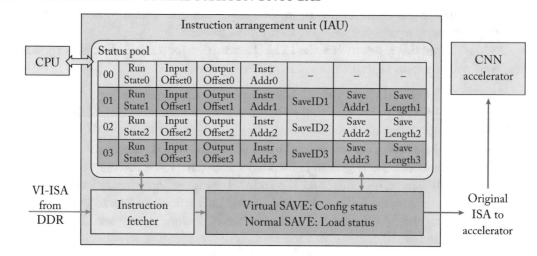

Figure 7.9: Hardware architecture of IAU. The software on the CPU (PS side) communicates with IAU to access the CNN accelerator. IAU records the running state of each task and translates the input instruction virtual instructions sequence (VI-ISA) to a normal sequence of instructions (Original ISA).

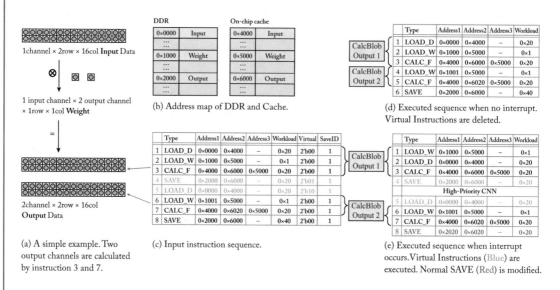

Figure 7.10: A simple example of our proposed virtual-instruction-based interrupt: (a) the CNN layer structure; (b) the on-chip and DDR addresses of different data; (c) the instruction sequence in virtual instruction ISA (VI-ISA); The blue instructions are the virtual SAVE and LOAD; (d) executed instructions when no interrupt occurs; and (e) executed instructions when an interrupt occurs.

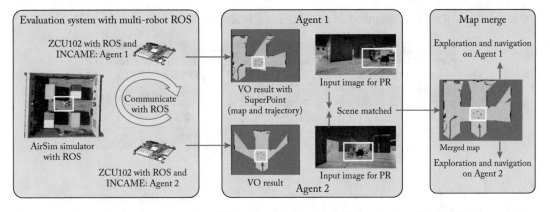

(a) Evaluation system based on multi-agent ROS. The AirSim simulator runs on the server, providing input images for two agents. The computation of each agent runs on a ZCU102 board with INCAME. The communication is based on ROS.

(b) The map and trajectory (left) are generated by VO based on SuperPoint feature-points. The two input pictures are from the same scene (right) with the chairs in white boxes. The PR representation of these two pictures is similar.

(c) After the same scene detected, the relative pose of the two agents is evaluated by the similar scene. The map and trajectory are merged via the relative pose.

Figure 7.11: Multi-robot exploration: environment and results: (a) the hardware-in-loop system based on ROS; (b) the robot components in Fig. 7.1, such as feature-point extraction (FE), place recognition (PR), visual odometery (VO) are processed on each hardware platform. The numbers in circle indicate the components in Fig. 7.1; and (c) the result of merged map and trajectories.

Thus, the two output channels are calculated by two CALC_F instructions (instructions 3 and 7 in Fig. 7.10c). The addresses used in the instruction example are listed in Fig. 7.10b. Figure 7.10c is the instruction sequence from DDR with VI-ISA. Figure 7.10d is the executed original ISA instructions without interrupt. When an interrupt occurs at the first CalcBlob, Fig. 7.10e illustrates the backup/recovery instructions (blue) and the modified SAVE instruction (Red).

7.4   EVALUATION AND RESULTS

7.4.1   EXPERIMENT SETUP

The hardware-in-the-loop evaluation environment is illustrated in Fig. 7.11a. There is a simulation server providing the simulation environment based on AirSim [331], which is a high-fidelity visual and physical simulation for autonomous vehicles. The AirSim simulation server provides the camera data for each agent. Two Xilinx ZCU102 boards [321], with ZU9 MPSoC [310], are responsible for the calculation of each agent. The components in Fig. 7.1 for each agent are implemented in the ZCU102 board. The implementation of the FE (① in Fig. 7.1), Super-

Table 7.2: Hardware consumption of the proposed hardware

	#DSP	#LUT	#FF	#BRAM
On-board resource	2520	274080	548160	912
CNN accelerator	1282	74569	171416	499
IAU	0	2268	4633	4
FE post-processing	25	17573	29115	10

Point [311]), is introduced in previous sections. GeM [72] is used to implement the PR module (②). GeM is a CNN-based method with ResNet101 as the backbone, and the post-processing of GeM calculates the 3-norm of the output feature maps.

The VO module (③) in the experiment is the PnP [332] method, which is widely used in the feature-point based VO. The DOpt module (④) is proposed in [308] and also used in the former distributed SLAM system [305]. The Map Merging (⑤), Exploration (⑥), and Navigation (⑦) in this work are provided by the ROS framework and running on the CPU side.

The hardware resources are listed in Table 7.2. The hardware resources are provided by Vivado after hardware implementation. Vivado is the development toolchain for MPSoC provided by Xilinx. The CNN backbone is calculated by the Angel-Eye CNN accelerator [92] on the FPGA side of ZCU102 (Programmable logic, PL side). The FE post-processing steps run on our proposed accelerators, also on the PL side. The PL side has two clock frequencies. The CNN accelerator and the IAU are running at 300 MHz. The accelerator for FE post-processing is running at 200 MHz. Compared with the CNN accelerator, IAU and FE post-processing use minimal hardware resources.

7.4.2   VIRTUAL INSTRUCTION-BASED INTERRUPTS

Interrupt Response Latency and Extra Cost

We evaluate the latency to respond the interrupt ($t_{latency}$) and the performance degradation (t_{cost}) of different interrupt method. In MR-Exploration, only the low-priority PR task is interruptible, and the interrupt position is unpredictable. GeM [72] is used to implement the PR module in the experiment. The CNN backbone of the GeM is ResNet101, which contains 101 convolution layers. The input shape of the CNN is $480 \times 640 \times 3$. The parallelism of the Angel-Eye is $P_h = 8$, $P_{in} = 16$, $P_{out} = 16$, i.e., each CALC instruction processes 8 lines from 16 input channels to 16 output channels.

The latency to respond the interrupt in CPU-like method consists of the time to finish current executing instruction and the data backup time ($t_{latency} = t_1 + t_2$) for the on-chip data/weights caches (totally 2.2 MB). The latency in layer-by-layer interrupt is the time to fin-

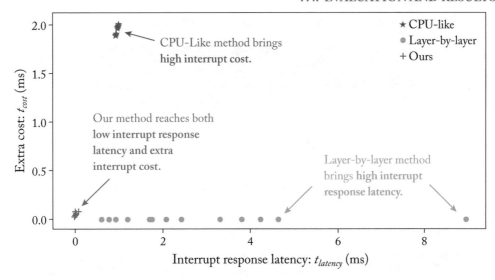

Figure 7.12: The interrupt response latency and extra time cost. Our *VI* method reaches both low latency and cost.

ish the current layer. The latency of our virtual-instruction-based method is the time to finish the currently executing instruction and the backup time for the calculated output results.

The cost of CPU-like interrupt is the data transfer time of all the on-chip caches (totally 2.2 MB) to/from DDR ($t_{cost} = t_2 + t_4$). The cost of our virtual-instruction-based method is only the recovery of the input/weights from DDR to on-chip caches ($t_{cost} = t_4$). There is no extra cost for the layer-by-layer interrupt.

We randomly sample 12 positions of the ResNet101 CNN backbone. The interrupt response latency and the extra time cost for different implementation of interrupt at the positions are listed in Fig. 7.12. The CPU-like interrupt consumes the most extra cost (t_{cost}). Though the layer-by-layer interrupt consumes no extra time, the latency is much higher than our virtual-instruction-based interrupt. This is because the layer-by-layer interrupt needs to wait for the completion of the whole layer. The performance at the same interrupt position in our proposed method can interrupt inside a layer with lower latency.

Furthermore, though the network structures differ between different CNNs, the convolutional layers, which are the basic component in CNN, are similar between different CNNs. INCAME monitors the running status inside each layer, and the interrupt response latency and extra cost are only relevant to the currently operating layer. Thus, the latency and cost are also similar between different CNNs. In conclusion, the process for different CNN tasks are similar, and the cost of different tasks are similar.

Table 7.3: Time comparison between data backup and calculation

H	W	Ch_{in}	Ch_{out}	Kernel Size	Backup (t_2,us)	Conv (t_1,us)	$\dfrac{Backup}{Conv}$
480	640	3	64	7 × 7	26.29	52.38	50.2%
120	160	128	128	3 × 3	8.77	41.18	21.3%
30	40	1024	2048	1 × 1	1.25	8.75	14.3%
30	40	512	512	3 × 3	1.42	39.36	3.6%
16	20	512	512	3 × 3	0.75	20.16	3.8%

Table 7.4: The instruction number of the interruptible PR network (ResNet101)

	Instr. Number	Instr. Volume (MB)	Execute Time (ms)
Original ISA	364032	4.36	186.0
VI-ISA	400243	4.80	186.4

Time Comparison Between t_1 and t_2

As described in Section 7.3.5, the layer-by-layer interrupt method do not need to backup data before interruption ($t_2 = 0$). Though our Virtual-Instruction method (*VI* method) need to spend time to backup the final results, which are already generated yet not stored to DDR (t_2). However, compared with computation, the time of data backup (t_2) is short. We list the backup time and the convolution time (t_1) at some of the interrupt positions in Fig. 7.12, with different feature map shapes, kernel sizes, and input/output channels. H, W are the height and width of input feature maps. Ch_{in}, Ch_{out} are the number of input and output channels. The time of backup and calculation is listed in *Backup* and *Conv* columns. The ratio of backup and calculation is listed in the last column. The backup operation only consumes less than 20% of the calculation time. For the first layer (first line of Table 7.3), the input channel number is too small, so the calculation time is also short, and thus the backup time has reached half of the calculation time.

Additional Data Transfer for the Virtual Instructions

The extra virtual instruction number is listed in Table 7.4. Compared to the normal instruction transfer, the volume of virtual instructions is less than 10%. The accelerator takes 186.0 ms to finish the original ISA sequence of GeM backbone, and only takes more 0.4 ms to fetch and translate the VI-ISA sequence without interruption. The performance degradation brought by the extra virtual instructions is negligible (less than 0.3%). We are here to remind the readers that the processing of the low-priority PR task can be interrupted twice or more. And in that case, the PR task is interrupted by several FE tasks.

7.4.3   ROS-BASED MR-EXPLORATION

The results of the Multi-Robot Exploration based on INCAME are shown in Fig. 7.11. The space in the AirSim [331] for the robots to explore is shown in Fig. 7.11a. It is a simple rectangle area with four different pillars and some chairs at the center (in the white box). Figure 7.11b shows how PR works for map merging. The FE and VO of each agent produce the local map and trajectory on each ZCU102 board. When the PR threads on different agents find out a similar scene, the relative pose of the two agents at the similar scene is calculated. The map and the trajectory are merged with the calculated relative pose, as shown in Fig. 7.11c.

In this example, the FE and PR are both executed on the same Angel-Eye accelerator. The frequency of the input camera is 20 FPS. As soon as an input frame is fed to the FE, and FE module would take up the accelerator. While the CPU process VO with the feature-points from FE, the accelerator can switch to process the low-priority PR task. Because the executing time of VO varies, the time to finish a PR task is different. In this example, the time from the beginning of a PR to its end is 320–500 ms. Thus, the PR process one keyframe every 7–10 input frames.

7.5   SUMMARY

In this chapter, we discussed how FPGAs can be utilized in multi-robot exploration tasks. Specifically, we have presented an FPGA-based interruptible CNN accelerator and a deployment framework, INCAME, for multi-robot exploration. With the help of the virtual-instruction-based interrupt method, the FPGA-based CNN accelerator can switch between different CNN tasks with low interrupt response latency and low extra cost. INCAME only needs to modify the instruction fetch module to IAU in hardware. Thus, it is easy to extend to handle other instruction-driven accelerators. Therefore, with the help of INCAME, the independent software in ROS can access the accelerator without hardware resource conflicts on various FPGA-based CNN accelerators. Note that the development of CPU task scheduling evolved from single-core multi-task to multi-core multi-task. Similarly, INCAME currently focuses on interrupt support for single-core multi-task. We plan to investigate the multi-core multi-tasking for FPGA-based CNN accelerators as part of future work. INCAME also accelerates the time-consuming post-processing operations like SoftMax, NMS, and normalization, so that the ROS-based MR-Exploration can achieve real-time performance on embedded FPGAs.

CHAPTER 8

Autonomous Vehicles Powered by FPGAs

FPGAs provide rich I/O interfaces, flexibility, and capability of handling complex workloads with high performance and low energy consumption, thus FPGAs are ideal compute substrates to deploy in autonomous driving systems. In this chapter, we present a detailed case study on building a commercial autonomous driving compute system, especially the choices between different compute units (mobile SoCs, GPUs, CPUs, FPGAs) and the utilization of FPGAs. Through this chapter, readers should be able to understand the importance of FPGAs in commercial robotic systems and the unique role FPGAs play in a heterogeneous compute system.

8.1   THE PERCEPTIN CASE STUDY

In 2017, PerceptIn decided to develop low-speed autonomous vehicles to serve the micromobility market, as micromobility is a rising transport mode wherein lightweight vehicles cover short trips that massive transit ignores [13]. According to the U.S. Department of Transportation, 60% of vehicle traffic is attributed to trips under 5 miles [333]. Transportation needs in short trips are disproportionally underserved by current mass transit systems due to high costs, which affects society profoundly. Micromobility bridges transit services and communities' needs, driving the rise of Mobility-as-a-Service.

Based on internal business analysis, if PerceptIn could provide low-speed autonomous vehicles under $70,000 per unit, PerceptIn would generate a reasonable return-on-investment for PerceptIn's customers, the autonomous vehicle operators. However, the $70,000 price tag also imposes very strict and challenging constraints on the design of low-speed autonomous vehicles. In detail, we have to break down the $70,000 into non-recurring engineering (NRE) cost such as research and development, recurring costs such as the cost of the chassis, the cost of drive-by-wire conversion (meaning to convert a traditional vehicle into one that can be controlled by computers), the cost of sensors, the cost of integration, the cost of customer service, and finally the cost of the computing system [334].

As the on-vehicle computing system is a significant contributor to cost and power consumption, PerceptIn conducted a study on autonomous driving computing systems [14, 16, 335], PerceptIn concluded that computing is the bottleneck for the commercial deployment of autonomous vehicles, and PerceptIn needed a computing system that is reliable, affordable, high-performance, and energy-efficient. Most importantly, PerceptIn needed a solution that is cost-effective and has a short time-to-market. PerceptIn faced several options.

Option one: Optimization of commercial off-the-shelf mobile System-on-Chips (SoCs). This approach brings several benefits, first, since mobile SoCs have reached economies of scale, it would have been the most beneficial for PerceptIn to build its technology stack on affordable, backward-compatible computing systems. Second, PerceptIn's vehicles target micromobility with limited speed, similar to mobile robots, for which mobile SoCs have been demonstrated before. However, an extensive study is required to fully understand mobile SoCs' suitability for autonomous driving, this may delay PerceptIn's product launch by six months.

Option two: Procurement of specialized autonomous driving compute systems. There were commercial compute platforms specialized for autonomous driving, such as those from NXP, MobilEye, and Nvidia. They are mostly Application-Specific Integrated Circuit (ASIC) based chips that provide high performance at a much higher cost. For instance, the first-generation of Nvidia PX2 system costs over $10,000. Besides the cost issue, these computing systems mostly accelerate only the perception function in autonomous driving, whereas PerceptIn requires a system that optimizes the end-to-end performance. So we soon concluded that this option was not viable.

Option three: Development of proprietary autonomous driving computing systems. Developing a proprietary computing system guarantees that PerceptIn has the most suitable system for PerceptIn's customers and its workloads, but also means that PerceptIn needs to invest a significant amount of financial and personnel resources on this project. Also, the investment does not guarantee the success of this project. It is a huge and risky bet for a startup like PerceptIn. After an unsuccessful exploration with option one, starting in early 2018, PerceptIn decided to move forward with option three and PerceptIn thus formed a team to develop the FPGA-based DragonFly computing system [263, 336–341]. Option three was a huge success, today all autonomous vehicles shipped by PerceptIn are empowered by PerceptIn's proprietary DragonFly compute system. In this chapter, from a technical perspective, we explained how we designed PerceptIn's DragonFly computing systems with FPGAs, and the design trade-offs that we faced.

8.2 DESIGN CONSTRAINTS

8.2.1 OVERVIEW OF THE VEHICLE

PerceptIn provides two autonomous vehicle designs: two-seater pods targeting private transportation experiences and eight-seater shuttles targeting public autonomous driving transportation services. Both designs are capped at 20 mph to suit the unique needs of micromobility. Our vehicles are fully autonomous without requiring human driver's intervention, complying with level 4 autonomous vehicle defined by the U.S. Department of Transportation's National Highway Traffic Safety Administration (NHTSA) [342].

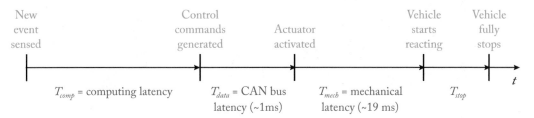

Figure 8.1: The end-to-end latency model. The computing latency, T_{comp}, is the main optimization target.

The unique scenarios and use-cases let us build commercial autonomous vehicles at a reasonable cost. The vehicles are sold at a price of $70,000, over $10 \times$ lower than what is commonly believed to be possible for commercial autonomous vehicles [9].

8.2.2   PERFORMANCE REQUIREMENTS

Latency Requirement   The end-to-end latency, i.e., the time between when a new event is sensed in the environment (e.g., change of distance, new object) and when the vehicle fully stops, must be short enough to avoid hitting objects.

The latency requirement could be derived using a simple analytical model, as shown in Fig. 8.1. The latency consists of four major components: the time for the computing system to generate control commands from the sensor inputs (T_{comp}), the time to transmit the control commands to the vehicle's actuator through the Controller Area Network (CAN) bus (T_{data}), the time it takes for the mechanical components of the vehicle to start reacting (T_{mech}), and the time for the vehicle to fully stop (T_{stop}). Assuming the brake generates a deceleration of a, the vehicle's speed is v, and an object is of distance D to the vehicle when it is sensed, the following, which is generic to any autonomous vehicle, must hold:

$$(T_{comp} + T_{data} + T_{mech}) \times v + \frac{1}{2} \times a \times T_{stop}^2 \leq D \qquad (8.1a)$$

$$T_{stop} = v/a. \qquad (8.1b)$$

We use this model to derive the latency requirement of the computing system T_{comp}, which is the primary target of our optimizations. In our vehicles at a typical speed v of 5.6 m/s, the brake generates a deceleration a of about 4 m/s². Our measurements show that T_{data} is about 1 ms and T_{mech} is about 19 ms. Given these, Fig. 8.2a shows that the computing latency requirement T_{comp} (y-axis) tightens as the object distance D (x-axis) becomes closer.

As the compute latency becomes lower, the vehicle could avoid objects that are farther away. Our vehicle has an average computing latency of 164 ms (Section 8.4.4), which means the vehicle could avoid any objects that are 5 m away or farther when they are detected. Our

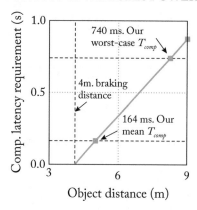
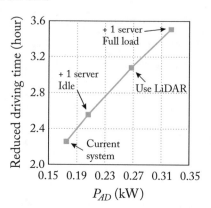

(a) The computing latency requirement becomes tighter when the object to be avoided is closer.

(b) Driving time reduces as the power of the autonomous driving system increases (P_{AD}).

Figure 8.2: Impacts of latency and energy.

worst-case computing latency is 740 ms, under which the vehicle could avoid objects that are detected at least 8.3 m away.

This latency model allows us to understand how much compute latency matters in the end-to-end system, which provides targets/guidances for hardware acceleration. For instance, as shown in Fig. 8.2a, if the vehicle is to plan the route to avoid an obstacle at 5 m away, the compute latency must be lower than 164 ms.

Throughput Requirement   The throughput requirement, quantified in the number of control commands sent to the actuator per second, dictates how often we can control the vehicle. Higher throughput allows more smooth control without abrupt turns and brakes. We set a 10 Hz throughput requirement for our vehicles, which is much higher than how often a human driver manipulates vehicles.

8.2.3   ENERGY AND COST CONSIDERATIONS

Energy Constraint and Impacts on Hardware   The computing systems and sensors that enable autonomous driving to consume extra energy, which reduces the driving range and translates to revenue loss for commercial vehicles. These considerations in turn impact the hardware design.

The impact of energy consumption on driving time reduction ($T_{reduced}$) could be derived using a simple model:

$$T_{reduced} = \frac{E}{P_V} - \frac{E}{P_V + P_{AD}}, \tag{8.2}$$

Table 8.1: Power breakdown of our vehicles. As a comparison, we also show the power consumption's of typical LiDARs, which are not used by us.

Component(s)		Power (W)	Quantity
Main computing	Dynamic	118	1
(CPU+GPU) server	Idle	31	1
Embedded vision module (FPGA + cameras/IMU/GPS)		11	1
Radars		13	6
Sonars		2	8
Total for AD (P_{AD})		175	–
Vehicle without AD (P_V)		600	–
LiDAR	Long-range [343]	60	1
(not used by us)	Short-range [343]	8	1

where P_V denotes the power consumption of the vehicle itself (without the autonomous driving capabilities),[1] P_{AD} denotes the additional power consumed to enable autonomous driving, and E denotes the vehicle's total energy capacity.

Our vehicles are electric cars that are powered by batteries, which have a total energy capacity of 6 kilowatt hour (kW·h). The vehicle itself consumes 0.6 kW on average (P_V),[2] and enabling autonomous driving consumes an additional 0.175 kW (P_{AD}). Effectively, supporting autonomous driving reduces the driving time on a single charge from 10–7.7 h.

Table 8.1 further breaks down the additional power consumption into four components (see Section 8.4.2 for a detailed hardware architecture): the main computing server (CPU + GPU), the embedded vision module (an FPGA board with the cameras, Inertial Measurement Unit (IMU), and GPS), the six Radar units, and the eight Sonar units. The computing server contributes the most to the additional power.

Energy consumption significantly impacts hardware system design. Using one of our deployments in a tourist site in Japan, let us consider the implication of adding one additional computing server—presumably to support advanced algorithms or to reduce computing latency. Since computing systems in autonomous vehicles are always-on when the vehicle is active, the idle power of the additional server alone would increase the total power consumption by 31 W, which reduces the driving time by 0.3 h. Each vehicle in that tourist site operates about 10 h a day; thus, the reduced operation time would translate to 3% revenue lost per day. Figure 8.2b

---

[1]Note that P_V is affected by the weight, which in turn consists of the weight of the vehicle itself and the weight of the passengers. The passenger weight in our two-seater car is about one-fifth of the vehicle itself. A detailed analytical model of P_V could be found at Kim et al. [344].

[2]The peak power could be as high as 2 kW.

Table 8.2: Cost breakdown of our vehicle and cost comparison with LiDAR-based vehicles

Vehicles	Components	Price
Our vehicle (camera-based)	Cameras × 4 + IMU	$1,000
	Radar × 6	$3,000
	Sonar × 8	$1,600
	GPS	$1,000
	Retail price	$70,000
LiDAR-based vehicle (e.g.,Waymo)	Long-range LiDAR	$80,000
	Short-range LiDAR × 4	$16,000
	Estimated retail price	>$300,000

shows how the driving time reduces as P_{AD} increases. If the additional server is operating at full load, the driving time is reduced by about 3.5 h.

Cost   In the long term, popularizing autonomous vehicles is possible only if they significantly reduce the cost of transportation services, which in turn imposes tight cost constraints on building and supporting the vehicle.

Similar to the concept of the total cost of ownership (TCO) in data centers [345], the cost of an autonomous vehicle is a complex function influenced by not only the cost of the vehicle itself, but also indirect costs such as servicing the vehicle and maintaining the back-end cloud services.

As a reference point, Table 8.2 provides a cost breakdown of our vehicles that operate in a tourist site in Japan. Considering all the factors influencing the vehicle's cost, each vehicle is sold at $70,000, which allows the tourist site to charge each passenger only $1 per trip. Any increase in the vehicle cost would directly increase the customer cost.

Cost comparison with LiDARs: The total cost of operating a vehicle that uses LiDAR (e.g., Waymo) is much higher. As shown in Table 8.2, a long-range LiDAR could cost about $80,000 [9], whereas our cameras + IMU setup costs about $1,000. The total cost of a LiDAR-equipped car could cost between $300,000 [346] to $800,000 [9]. In addition, creating and maintaining the High-Definition (HD) maps required for LiDARs can cost millions of dollars per year for a mid-size city [347] due to the need for high resolution.

8.3   SOFTWARE PIPELINE

Figure 8.3 shows the block diagram of our on-vehicle processing system, which consists of three components: sensing, perception, and planning. Sensor samples are synchronized and processed before being used by the perception module, which performs two fundamental tasks: (1) understanding the vehicle itself by localizing the vehicle in the global map (i.e., ego-motion esti-

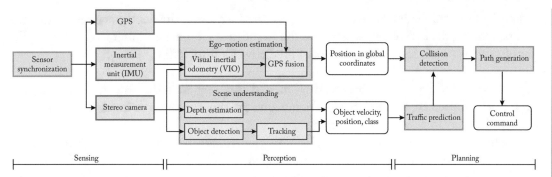

Figure 8.3: Block diagram of our on-vehicle processing software system.

Table 8.3: Algorithms explored in perception and planning

Task	Algorithm(s)	Sensors(s)
Depth estimation	ELAS [352]	Cameras
Object detection	YOLO [354]/Mask R-CNN [355]	Camera
Object tracking	KCF [353]	Camera
Localization	VIO [349]	Cameras, IMU, GPS
Planning	MPC [216]	–

mation) and (2) understanding the surroundings through depth estimation and object detection/tracking. The perception results are used by the planning module to plan the path and to generate the control commands. The control commands are transmitted through the CAN bus to the vehicle's Engine Control Unit (ECU), which then activates the vehicle's actuator, from which point on the vehicle's mechanical components take over.

Algorithms Table 8.3 summarizes the main algorithms explored in our computing system. The localization module is based on the classic Visual Inertial Odometry algorithm [348, 349], which uses camera-captured images and IMU samples to estimate the vehicle's position in the global map. Depth estimation uses stereo vision algorithms, which calculate object depth by processing the two slightly different images captured from a stereo camera pair [350, 351]. In particular, ELAS algorithm, which uses hand-crafted features [352], is used. While DNN models for depth estimation exist, they are orders of magnitude more compute-intensive than non-DNN algorithms [246] while providing marginal accuracy improvements to our use-cases.

Object detection is achieved by using DNN models. Object detection is the only task in the current pipeline where the accuracy provided by deep learning justifies the overhead. An object, once detected, is tracked across time until the next set of detected objects are available. The DNN models are trained regularly using our field data. As the deployment environment can vary significantly, different models are specialized/trained using the deployment environment-specific training data. Tracking is based on the Kernelized Correlation Filter (KCF) [353]. The planning algorithm is formulated as Model Predictive Control (MPC) [216].

Task-Level Parallelism   Sensing, perception, and planning are serialized; they are all on the critical path of the end-to-end latency. We pipeline the three modules to improve the throughput, which is dictated by the slowest stage.

Different sensor processings (e.g., IMU vs. camera) are independent. Within perception, localization and scene understanding are independent and could execute in parallel. While there are multiple tasks within scene understanding, they are mostly independent with the only exception that object tracking must be serialized with object detection. The task-level parallelisms influence how the tasks are mapped to the hardware platform.

8.4   ON VEHICLE PROCESSING SYSTEM

This section provides a detailed description of the on-vehicle processing system, which we call "Systems-on-a-Vehicle" (SoV).

8.4.1   HARDWARE DESIGN SPACE EXPLORATION

Before introducing our hardware architecture for on-vehicle processing, this section provides a retrospective discussion of the two hardware solutions we explored in the past but decided not to adopt: off-the-shelf mobile SoCs and off-the-shelf (ASIC-based) automotive chips. They are at the two extreme ends of the hardware spectrum; our current design combines the best of both worlds.

Mobile SoCs   We initially explored the possibility of utilizing high-end mobile SoCs, such as Qualcomm Snapdragon [356] and Nvidia Tegra [357], to support our autonomous driving workloads. The incentive is two-fold. First, since mobile SoCs have reached economies of scale, it would have been most beneficial for us to build our technology stack on affordable, backward-compatible computing systems. Second, our vehicles target micromobility with limited speed, similar to mobile robots, for which mobile SoCs have been demonstrated before [16, 358–360].

However, we found that mobile SoCs are ill-suited for autonomous driving for three reasons. First, the compute capability of mobile SoCs is too low for realistic end-to-end autonomous driving workloads. Figure 8.4 shows the latencies and energy consumptions of three perception tasks—depth estimation, object detection, and localization—on an Intel Coffee Lake CPU (3.0 GHz, 9 MB LLC), Nvidia GTX 1060 GPU, and Nvidia TX2 [361]. Figure 8.4a shows that TX2 is much slower than the GPU, leading to a cumulative latency of 844.2 ms for perception alone. Figure 8.4 shows that TX2 has only marginal, sometimes even worse, energy reduction compared to the GPU due to the long latency.

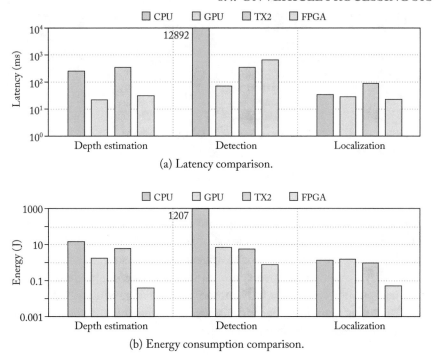

(a) Latency comparison.

(b) Energy consumption comparison.

Figure 8.4: Performance and energy comparison of four different platforms running three perception tasks. On TX2, we use the Pascal GPU for depth estimation and object detection and use the ARM Cortex-A57 CPU (with SIMD capabilities) for localization, which is ill-suited for GPU due to the lack of massive parallelisms.

Second, mobile SoCs do not optimize data communication between different computing units, but require redundant data copying coordinated by the power-hungry CPU. For instance, when using DSP to accelerate image processing, the CPU has to explicitly copy images from sensor interface to DSP through the entire memory hierarchy [338, 362, 363].

Finally, traditional mobile SoCs design emphasizes compute optimizations, while we find for autonomous vehicle workloads, sensor processing support in hardware is equally important. For instance, autonomous vehicles require very precise and clean sensor synchronization, which mobile SoCs do not provide.

Automotive ASICs At the other extreme are computing platforms specialized for autonomous driving, such as those from NXP [364], MobileEye [365], and Nvidia [366]. They are mostly ASIC-based chips that provide high performance at a much higher cost. For instance, the first-generation of Nvidia PX2 system costs over $10,000 while an Nvidia TX2 SoC costs only $600.

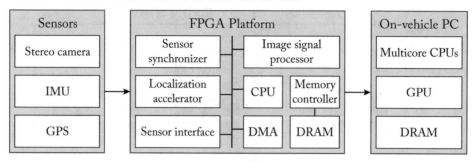

Figure 8.5: Overview of our SoV hardware architecture.

Besides the cost, we do not use existing automotive platforms for two technological reasons. First, they focus on accelerating a subset of the tasks in fully autonomous driving. For instance, Mobileye chips provide assistance to drivers through features such as forward collision warnings. Thus, they accelerate key perception tasks (e.g., lane/object detection). Our goal is an end-to-end computing pipeline for fully autonomous driving, which must optimize sensing, perception, and planning as a whole. Second, similar to mobile SoCs, these automotive solutions do not provide good sensor synchronization support.

8.4.2　HARDWARE ARCHITECTURE

The results from our hardware exploration motivate the current SoV design. Our design (1) uses an on-vehicle server machine to provide sufficient compute capabilities, (2) uses an FPGA platform to directly interface with sensors to reduce redundant data movement while providing hardware acceleration for key tasks, and (3) uses software-hardware collaborative techniques for efficient sensor synchronization.

Overview

Figure 8.5 shows an overview of the SoV computing system. It consists of sensors and server+FPGA heterogeneous computing platforms.

The sensing hardware consists of (stereo) cameras, IMU, and GPS. In particular, our vehicle is equipped with two sets of stereo cameras, one forward facing and the other backward facing, for depth estimation. One of the cameras in each stereo pair is also used to drive monocular vision tasks such as object detection. The cameras along with the IMU drive the VIO-based localization algorithm.

Considering the cost, compute requirements, and power budget, our current computing platform consists of a Xilinx Zynq UltraScale+FPGA board and an on-vehicle PC machine equipped with an Intel Coffee Lake CPU and an Nvidia GTX 1060 GPU. The PC is the main computing platform, but the FPGA plays a critical role, which bridges sensors and the server and provides an acceleration platform.

Algorithm-Hardware Mapping

There is a large design space in mapping the sensing, perception, and planning tasks to the FPGA + server (CPU + GPU) platform. The optimal mapping depends on a range of constraints and objectives: (1) the performance requirements of different tasks, (2) the task-level parallelism across different tasks, (3) the energy/cost constraints of our vehicle, and (4) the ease of practical development and deployment.

Sensing   We map sensing to the Zynq FPGA platform, which essentially acts as a sensor hub. It processes sensor data and transfers sensor data to the PC for subsequent processing. The Zynq FPGA hosts an ARM-based SoC and runs a full-fledged Linux OS, on which we develop sensor processing and data transfer pipelines.

The reason sensing is mapped to the FPGA is three-fold. First, embedded FPGA platforms today are built with rich/mature sensor interfaces (e.g., standard MIPI Camera Serial Interface [367]) and sensor processing hardware (e.g., ISP [368]) that server machines and high-end FPGAs lack.

Second, by having the FPGA directly process the sensor data in situ, we allow accelerators on the FPGA to directly manipulate sensor data without involving the power-hungry CPU for data movement and task coordination.

Finally, processing sensor data on the FPGA naturally let us design a hardware-assisted sensor synchronization mechanism, which is critical to sensor fusion.

Planning   We assign the planning tasks to the CPU of the on-vehicle server, for two reasons. First, the planning commands are sent to the vehicle through the CAN bus; the CAN bus interface is simply more mature on high-end servers than embedded FPGAs. Executing the planning module on the server greatly eases deployment. Second, as we will show later (Section 8.4.4) planning is very lightweight, contributing to about 1% of the end-to-end latency. Thus, accelerating planning on FPGA/GPU has marginal improvement.

Perception   Perception tasks include scene understanding (depth estimation and object detection/tracking) and localization, which are independent and, therefore, the slower one dictates the overall perception latency.

Figure 8.4a compares the latency of the perception tasks on the embedded FPGA with the GPU. Due to the available resources, the embedded FPGA is faster than the GPU only for localization, which is more lightweight than other tasks. Since latency dictates user experience and safety, our design offloads localization to the FPGA while leaving other perception tasks on the GPU. Overall, the localization accelerator on FPGA consumes about 200 K LUTs, 120 K registers, 600 BRAMs, 800 DSPs, with less than 6 W power.

This partitioning also frees more GPU resources for depth estimation and object detection, further reducing latency. Figure 8.6 compares different mapping strategies. When both scene understanding and localization execute on the GPU, they compete for resources and slow down each other. Scene understanding takes 120 ms and dictates the perception latency. When

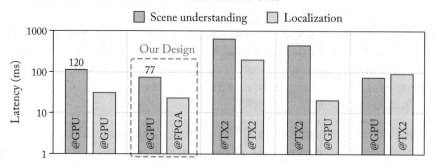

Figure 8.6: Latency comparison of different mapping strategies of the perception module. The perception latency is dictated by the slower task between scene understanding (depth estimation and object detection/tracking) and localization.

localization is offloaded to the FPGA, the localization latency is reduced from 31–24 ms, and scene understanding's latency reduces to 77 ms. Overall, the perception latency improves by 1.6 ×, translating to about 23% end-to-end latency reduction on average.

Figure 8.6 also shows the performance when using an off-the-shelf Nvidia TX2 module instead of an FPGA. While significantly easier to deploy algorithms, TX2 is always a latency bottleneck, limiting the overall computing latency.

8.4.3 SENSOR SYNCHRONIZATION

Sensor synchronization is crucial to perception algorithms, which fuse multiple sensors. For instance, vehicle localization uses Visual-Inertial Odometry, which uses both camera and IMU sensors. Similarly, object depth estimation uses stereo vision algorithms, which require the two cameras in a stereo system to be precisely synchronized.

An ideal sensor synchronization ensures that two sensor samples that have the same timestamp capture the same event. This in turn translates to two requirements. First, all sensors are triggered simultaneously. Second, each sensor sample is associated with a precise timestamp.

Out-of-sync sensor data is detrimental to perception. Figure 8.7a shows how unsynchronized stereo cameras affect depth estimation in our perception pipeline. As the temporal offset between the two cameras increases (further to the left on x-axis), the depth estimation error increases (y-axis). Even if the two cameras are off-sync by only 30 ms, the depth estimation error could be as much as 5 m. Figure 8.7b shows how unsynchronized camera and IMU would affect the localization accuracy. When the IMU and camera are off by 40 ms, the localization error could be as much as 10 m.

This section presents a software-hardware collaborative sensor synchronization design on FPGA. Figure 8.8 shows the high-level diagram of the system, which is based on two principles: (1) trigger sensors simultaneously using a single common timing source; and (2) obtain each

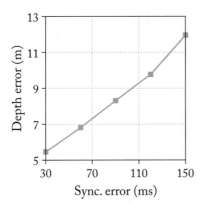

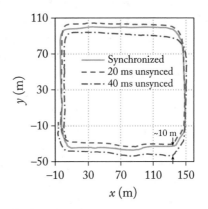

(a) Depth estimation error increases as stereo cameras become more out-of-sync. Even if the two cameras are off by only 30 ms, the depth estimation error could be over 5 m.

(b) Localized trajectory comparison between using synchronized and unsynchronized (camera and IMU) sensor data. The localization error could be as high as 10 m.

Figure 8.7: Importance of sensor synchronization.

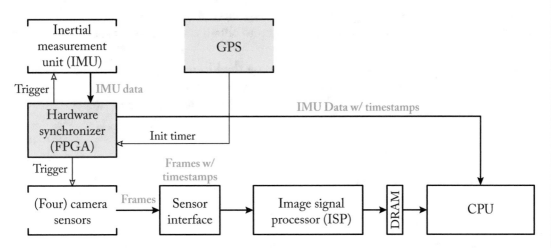

Figure 8.8: Sensor synchronization architecture. GPS provides a single common timing source to trigger cameras and IMU simultaneously. IMU timestamps are obtained in the hardware synchronizer, whereas camera timestamps are obtained in the sensor interface and are later adjusted in software by compensating for the constant delay between the camera sensor and the sensor interface.

sensor sample's timestamp that accurately captures the sensor triggering time. Guided by these two principles, a hardware synchronizer is designed, which triggers the camera sensors and the IMU using a common timer initialized by the satellite atomic time provided by the GPS device. The camera operates at 30 FPS while the IMU operates at 240 FPS. Thus, the camera triggering signal is downsampled 8 times from the IMU triggering signal, which also guarantees that each camera sample is always associated with an IMU sample.

Equally important to triggering sensors simultaneously is to pack a precise timestamp with each sensor sample. This must be done by the synchronizer as sensors are not built to do so. A hardware-only solution would be for the synchronizer to first record the triggering time of each sensor sample, and then pack the timestamp with the raw sensor data before sending the timestamped sample to the CPU. This is indeed what is done for the IMU data as each IMU sample is very small in size (20 Bytes). However, applying the same strategy to cameras would be grossly inefficient, because each camera frame is large in size (e.g., about 6 MB for a 1080p frame). Transferring each camera frame to the synchronizer just to pack a timestamp is inefficient.

Instead, our design uses a software-hardware collaborative approach shown in Fig. 8.8, where camera frames are directly sent to the SoC's sensor interface without the timestamps from the hardware synchronizer. The sensor interface then timestamps each frame. Compared with the exact camera sensor trigger time, the moment that a frame reaches the sensor interface is delayed by the camera exposure time and the image transmission time. These delays could be derived from the camera sensor specification. Applications can adjust the timestamp by subtracting the constant delay from the packed timestamp to obtain the triggering time of the camera sensor.

The hardware synchronizer is implemented on the FPGA. It is extremely lightweight in design with only 1,443 LUTs and 1,587 registers and consumes 5 mW of power. The synchronization solution is also fast. It incurs less than 1 ms delay to the end-to-end latency.

8.4.4 PERFORMANCE CHARACTERIZATIONS

This section characterizes the end-to-end computing latency on our hardware platform.

Average Latency and Variation Figure 8.9a breaks down the computing latency into sensing, perception, and planning. We show the best- and average-case latency as well as the 99th percentile. The mean latency (164 ms) is close to the best-case latency (149 ms), but a long tail exists.

To showcase the variation, the median latency of localization is 25 ms and the standard deviation is 14 ms. The variation is mostly caused by the varying scene complexity. In dynamic scenes (large change in consecutive frames), new features can be extracted in every frame, which slows down the localization algorithm.

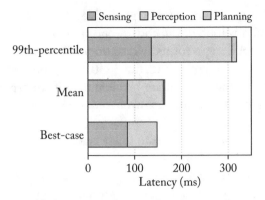
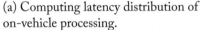

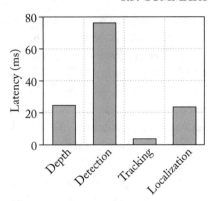

(a) Computing latency distribution of on-vehicle processing.

(b) Average-case latencies of different perception tasks.

Figure 8.9: Latency characterizations.

Latency Distribution   Unlike conventional wisdom, sensing contributes significantly to the latency. The sensing latency is bottlenecked by the camera sensing pipeline, which in turn is dominated by the image processing stack on the FPGA's embedded SoC (e.g., ISP and the kernel/driver), suggesting a critical bottleneck for improvement.

Apart from sensing, perception is the biggest latency contributor. Object detection (DNN) dominates the perception latency. Figure 8.9b further shows the latencies of different perception tasks in the average case. Recall that all perception tasks are independent except that object detection and tracking are serialized. Therefore, the cumulative latency of detection and tracking dictates the perception latency.

Planning is relatively insignificant, contributing to only 3 ms in the average case. This is because our vehicles require only coarser-grained motion planning at a lane granularity. Fine-grained motion planning and control at centimeter granularities [216], usually found in mobile robots [285], unmanned aerial vehicles [369], and many LiDAR-based vehicles, are much more compute-intensive. One such example is the Baidu Apollo EM Motion Planner [370], whose motion plan is generated through a combination of Quadratic Programming (QP) and Dynamic Programming (DP).

The sensing, perception, and planning modules operate at 10–30 Hz, meeting the throughput requirement.

8.5   SUMMARY

In this chapter, we have provided a retrospective summary of PerceptIn's efforts on developing on-vehicle computing systems for autonomous vehicles. It shows that autonomous driving incorporates a myriad of different tasks across computing and sensing domains with new design constraints. While accelerating individual algorithms is extremely valuable, what eventually matters is the systematic understanding and optimization of the end-to-end system.

Many related works have studied accelerator designs for individual on-vehicle processing tasks such as localization [371, 372], DNN [373–377], depth estimation [246, 378], and motion planning [285, 369], but meaningful gains at the system level are possible only if we expand beyond optimizing individual accelerators to exploiting the interactions across accelerators, a.k.a. accelerator-level parallelism [379, 380].

While the most common form of accelerator-level parallelism today is found on a single chip (e.g., in a smartphone SoC), accelerator-level parallelism in autonomous vehicles usually exists across multiple chips. For instance, in PerceptIn's computing platform, localization is accelerated on an FPGA while depth estimation and object detection are accelerated by a GPU. This case study has demonstrated that FPGAs are capable of playing a crucial role in autonomous driving, and exploiting accelerator-level parallelism while taking into account constraints arising in different contexts could significantly improve on-vehicle processing.

CHAPTER 9

Space Robots Powered by FPGAs

Due to radiation tolerance requirements, the compute power of space-grade ASICs is usually decades behind the state-of-the-art commercial off-the-shelf processors. On the other hand, space-grade FPGAs deliver high reliability, adaptability, processing power, and energy efficiency, and are expected to close the two-decade performance gap between commercial processors and space-grade ASICs, especially when it comes to powering space exploration robots. In the past 20 years, because of the advantages delivered by space-grade FPGAs, we have observed an increasing demand for the utilization of FPGAs in space robotic applications [381]. In this chapter, we introduce radiation tolerance for space computing, space robotic algorithm acceleration on FPGAs, as well as utilization of FPGAs in space robotic missions. Especially with the recent boom of commercial space exploration, we expect space robotic application will become the next demand driver for FPGAs.

9.1   RADIATION TOLERANCE FOR SPACE COMPUTING

For electronics intended to operate in space, the harsh space radiation present is an essential factor to consider. Radiation has various effects on electronics, but the commonly focused two are total ionizing dose effect (TID) and single event effects (SEE). TID results from the accumulation of ionizing radiation over time, which causes permanent damage by creating electron-hole pairs in the silicon dioxide layers of MOS devices. The effect of TID is that electronics gradually degrade in their performance parameters and eventually fail to function. Electronics intended for application in space are tested for the total amount of radiation, measured in kRads, they can endure before failure. Usually, electronics that can withstand 100 kRads are sufficient for low earth orbit missions to use for several years [52].

SEE occurs when high-energy particles from space radiation strike electronics and leave behind an ionized trail. The results are various types of SEEs [382] which can be categorized as either soft errors, which usually do not cause permanent damage, or hard errors, which often cause permanent damage. Examples of soft error include single event upset (SEU) and single event transient (SET). In SEU, a radiation particle struck a memory element, causing a bit flip. Noteworthy is that as the cell density and clock rate of modern devices increases, multiple cell upset (MCU), corruption of two or more memory cells in a single particle strike, is increasingly becoming a concern. A special type of SEU is single event functional interrupt (SEFI), where the upset leads to loss of normal function of the device by affecting control registers or the clock. In SET, a radiation particle passes through a sensitive node, which generates a transient voltage pulse, causing the wrong logic state at the combinatorial logic output. Depending on whether the impact occurs during an active clock edge or not, the error may or may not propagate. Some examples of hard error include single event latch-up (SEL), in which energized particle activates parasitic transistor and then cause a short across the device, and single event burnout (SEB), in which radiation induces high local power dissipation, leading to device failure. In these hard error cases, radiation effects may cause the failure of an entire space mission.

Space-grade FPGAs can withstand considerable levels of TID and have been designed against most destructive SEEs [383]. However, SEU susceptibility is pervasive. For the most part, radiation effects on FPGAs are not different from those of other CMOS-based ICs. The primary anomaly stems from FPGAs' unique structure, involving programmable interconnections. Depending on their type, FPGAs have different susceptibility toward SEU in their configuration. SRAM FPGAs are designated by NASA as the most susceptible ones due to their volatile nature. Even after the radiation hardening process, the configuration of SRAM FPGAs is only designated as "hardened" or simply having embedded SEE mitigation techniques rather than "hard," which means close to immune [52]. Configuration SRAM is not used in the same way as the traditional SRAM. A bit flip in configuration causes an instantaneous effect without the need for a read-write cycle. Moreover, instead of producing one single error in the output, the bit flip shifts the user logic directly, changing the device's behavior. Scrubbing is needed to rectify SRAM configuration. Antifuse and flash FPGAs are less susceptible to effects in configuration and are designated "hard" against SEEs in their configuration without applying radiation hardening techniques [52].

Design-based SEU/fault mitigation techniques are commonly used, for, in contrast to fabrication level radiation hardening techniques, they can be readily applied to commercial off-the-shelf (COTS) FPGAs. These techniques can be classified into static and dynamic. Static techniques rely on fault-masking, toleration of error without requiring active fixing. One such example is passive redundancy with voting mechanisms. Dynamic techniques, in contrast, detect faults and act to correct them. The common SEU Mitigation Methods include [384, 385] the following.

1. **Hardware Redundancy**: functional blocks are replicated to detect/tolerate faults. Triple modular redundancy (TMR) is perhaps the most widely used mitigation technique. It can be applied to entire processors or parts of circuits. At a circuit level, registers are implemented using three or more flip-flops or latches. Then, voters compare the values and output of the majority, reducing the likelihood of error due to SEU. As internal voters are also susceptible to SEU, they are sometimes triplicated also. For mission-critical applications, global signals may be triplicated to mitigate SEUs further. TMR can be implemented at ease with the help of supporting HDLs [386]. It is important to note that a limitation of

TMR is that one fault, at most, can be tolerated per voter stage. As a result, TMR is often used with other techniques, such as scrubbing, to prevent error accumulation.

2. **Scrubbing**: The vast majority of memory cells in reprogrammable FPGAs contain configuration information. As discussed earlier, configuration memory upset may lead to alteration routing network, loss of function, and other critical effects. Scrubbing, refreshing, and restoration of configuration memory to a known-good state, is therefore needed [385]. The reference configuration memory is usually stored in radiation-hardened memory cells either off or on the device. Scrubbers, processors, or configuration controllers carry out scrubbing. Some advanced SRAM FPGAs, including ones made by Xilinx, support partial reconfiguration, which allows memory repairs to be made without interrupting the operation of the whole device. Scrubbing can be done in frame-level (partial) or device-level (full), which will inevitably lead to some downtime; some devices may not be able to tolerate such an interruption. Blind scrubbing is the most straightforward way of implementation: individual frames are scrubbed periodically without error detection. Blind scrubbing avoids the complexity required in error detection, but extra scrubbing may increase vulnerability to SEUs as errors may be written into frames during the scrubbing process. An alternative to blind scrubbing is readback scrubbing, where scrubbers actively detect errors in configuration through error-correcting code or cyclic redundancy check [384]. If an error is found, scrubber initiates frame-level scrubbing.

Currently, the majority of space-grade FPGA comes from Xilinx and Microsemi. Xilinx offers the Virtex family and Kintex. Both are SRAM-based, which have high flexibility. Microsemi offers antifuse based RTAX and Flash-based RTG4, RT PolarFire, which have lower susceptibility against SEE and power consumption. 20 nm Kintex and 28 nm RT PolarFire are the latest generations. The European market is offered with Atmel devices and NanoXplore space-grade FPGAs [387]. Table 9.1 shows the specifications of the above devices.

9.2 SPACE ROBOTIC ALGORITHM ACCELERATION ON FPGAS

Robots in space rely heavily on computationally intensive algorithms to operate. The performance of current space-grade CPUs, lags decades behind that of the commercial parts, achieving only 22-400 MIPS [388] while a modern Intel i7 can easily achieve 200,000 MIPS. The insufficient performance of space-grade CPUs causes a bottleneck in robots' performance. In those circumstances, high-performance space FPGAs may be able to serve as accelerators, enhancing the performance of space robots. By implementing FPGA co-processors, suitable computationally expensive tasks can be offloaded onto FPGA co-processors for acceleration as FPGAs allow massive parallelization. Furthermore, FPGAs are generally more power-efficient, and they offer unique run-time reconfiguration ability allowing for their multipurpose use.

Table 9.1: Specifications of space-grade FPGAs

Device	Logic	Memory	DSPs	Technology	Radiation Tolerance
Xilinx Virtex-5QV	81.9K LUT6	12.3 Mb	320	65 nm SRAM	SEE immune up to LET>100 MeV/(mg·cm^2) and 1 Mrad TID
Xilinx RT Kintex UltraScale	331K LUT6	38 Mb	2760	20 nm SRAM	SEE immune up to LET>80 MeV/(mg·cm^2) and 100–120 Krads TID
Microsemi RTG4	150K LE	5 Mb	462	65 nm Flash	SEE immune up to LET>37 MeV(mg·cm^2) and TID>100 Krads
Microsemi RT PolarFire	481K LE	33 Mb	1480	28 nm Flash	SEE immune up to LET>63 MeV(mg·cm^2) and 300 Krads
Microsemi RTAX	4M gates	0.5 Mb	120	150 nm antifuse	SEE immune up to LET>37 MeV(mg·cm^2) and 300 Krads TID
Atmel ATFEE560	560K gates	0.23 Mb	–	180 nm SRAM	SEL immune up to 95 MeV(mg·cm^2) and 60 Krads TID
NanoXplore NG-LARGE	137K LUT4	9.2 Mb	384	65 nm SRAM	SEL immune up to 60 MeV(mg·cm^2) and 100 Krads TID

9.2.1   FEATURE DETECTION AND MATCHING

For Mars rover, one of the most exhausting tasks is localization and navigation, which utilizes various computer vision (CV) algorithms. Low performance in space-grade CPUs leads to the long execution time of those algorithms, making it the main factor of rovers' slow speed (5 cm/s). In localization and navigation of the Mars rovers, image feature detection and matching is of special importance. They are used in various CV tasks such as structure-from-motion, the process of constructing a 3D structure from a set of 2D images, and visual odometry, the method of incrementally determining a rover's pose by comparing sets of images taken by the rover over time.

In feature detection and matching, salient (keypoint) features in images are justified. Surrounding geometric properties are recorded by feature descriptors such that the salient features are easily identified and matched from multiple viewpoints and under various viewing conditions, achieving a high degree of repeatability. The detected features from a set of images are then matched against each other to establish a correspondence between the images. Some of the commonly used algorithms for this part are Harris detector [389], Feature from Accelerated Segment Test (FAST) [390], Scale-Invariant Feature Transform (SIFT) [391], Speed Up Robust Feature (SURF) [392], and Binary Robust Independent Elementary Features (BRIEF) [249]. Often, the features extracted from an image are in the thousands range, resulting in computationally intensive calculations that space-grade CPUs take too long to execute. However, the implementation of FPGA accelerators can reduce the execution time by one to two orders of magnitude as it enables massive parallel processing. The result of an experiment [393] may contextualize the dramatic acceleration. In the experiment, the above algorithms processed 512×384 images with 1000 features on a 150 MIPS space-grade CPU equivalence. Accelerators were implemented on a Virtex-6 FPGA (XC6VLX240T), which has 150 k LUT and 768 DSP slices, with 64x parallelization.

9.2.2   STEREO VISION

Planetary rovers need to reconstruct their surroundings to operate, no matter if it is for simultaneous localization and mapping (SLAM) or risk assessment involved in route selection. With rovers usually equipped with a pair of stereo cameras, the 3D surrounding environment reconstruction can be achieved through stereo vision algorithms. The algorithms correspond pixels from the stereo image pair and find the disparity between matched areas to perform calculation, such as triangulation, to obtain the depth of each pixel (dense stereo). Stereo vision algorithms, especially when dealing with high-resolution images, are computationally demanding. In a study conducted [394], it is shown that by implementing a Virtex-5 FPGA co-processor, which has 40960 slices and 576 BRAM, the processing time of stereo vision algorithms decreased significantly compared to using only space-grade CPU. When processing a pair of stereo images with 256 pixels in width, the run time using RAD6000, which runs at 20 MHz and processes 22 MIPS, was between 24–30 s. Processing time is reduced to around 1.32–1.65 s when using the newer RAD750 CPU, which processes 400 MIPS. After implementing Virtex-5 FPGA as the co-processor, the run time decreased to 0.005 s, more than 300x speed up from the RAD750 CPU and around 5000x from the RAD6000 CPU. Meanwhile, the algorithms take up 43% of slices and 12% of BRAM available.

9.2.3   DEEP LEARNING

FPGA acceleration in space missions goes beyond computer vision. Applying deep learning to space missions can increase the mission's level of autonomy and thus gaining efficiency. Reinforcement learning is a type of learning algorithm that determines what action an agent should take in a random environment to maximize future reward based on trial-and-error interactions with the environment [395]. The capabilities of this type of learning can be boosted by the integration of deep learning neural networks. However, the implantation of deep learning algorithms requires immense processing capability and execution time, which render it impractical to use on many space missions. However, FPGA implementations of such algorithms may enable their applications in space. In a study [396], Q-learning, a type of reinforcement learning, with artificial neural networks was developed, and its implementation on a Virtex-7 FPGA (XC7VX485T), which has 303 K LUTs and 2800 DSP Slices, results in a 30% power-saving and 43x speedup compared to an i5 6th Gen 2.3 GHz processor, again verifying FPGAs overwhelming advantages over typical space-grade processors.

The achievements that the world has made in deep learning, computer vision, and other related technology fields can be of great benefit to future space missions, increasing their level of autonomy and capabilities. Good onboard processors, ones that have high reliability, adaptability, processing power, and power efficiency, are needed in order to implement those technologies practically. The properties of FPGAs make them suitable for onboard processors: they have been used for space missions for decades and are proven in reliability; they have unrivaled adaptability and can even be reconfigured in run time; their capability for high degree parallel processing allow significant acceleration in executing many complex algorithms; hardware/software co-design methods make them potentially more power-efficient. Even though FPGAs significantly speed up algorithms compared to space-grade CPUs, the space-grade FPGAs are still generations behind their commercial off-the-shelf (COTS) counterparts. For example, at the time when the Virtex-7 series came out in the COTS version, the space-grade ones were still Virtex-5, two generations behind. In addition, the COTS version carries a lower price tag. As the radiation hardening requirement differs from mission to mission, COTS parts complemented with SEE mitigation techniques might be applicable to some space applications with low-reliability requirements or in non-mission critical parts. With the benefit of COTS FPGAs, their use in space might be the next step in the industry.

9.3   UTILIZATION OF FPGAS IN SPACE ROBOTIC MISSIONS

For space robotics, processing power is of particular importance, given the range of information required to accurately and efficiently process. Many of the current and previous space missions are packed with sophisticated algorithms that are mostly static. They serve to increase the efficiency of data transmission; nevertheless, data processing is done mainly on the ground. As the travel distance of missions increases, transmitting all data to, and processing it on the ground is no longer an efficient or even viable option due to transmission delay. As a result, space robots need to become more adaptable and autonomous. They will also need to pre-process a large amount of collected data and compress it before sending back to Earth [397].

The rapid development of the new generation of FPGAs may fill the need in space robotics. FPGAs enable robotic systems to be reconfigurable in real-time, making the systems more adaptable by allowing them to respond more efficiently to changes in environment and data. As a result, autonomous reconfiguration and performance optimization can be achieved simultaneously. Also, the FPGAs have a high capability for parallel processing, which is useful

Figure 9.1: Mars exploration rover.

in boosting processing performance. The use of FPGAs is present in various space robots, including some of the most prominent examples of the application, are the NASA Mars rovers. Since the first pair of rovers were launched in 2003, the presence of FPGAs has steadily increased in the later rovers.

9.3.1   MARS EXPLORATION ROVER MISSIONS

Beginning in the early 2000s, NASA has been using FPGAs in exploration rover control and lander control. In Opportunity and Spirit, the two Mars rovers launched in 2003, two Xilinx Virtex XQVR1000s were in the motor control board [398], which operates motors on instruments as well as rover wheels. In addition, an Actel RT 1280 FPGA was used in each of the 20 cameras on the rovers to receive and dispatch hardware commands. The camera electronics consist of a clock driver that provides timing pulses through the charge-coupled device (CCD), an IC containing an array of linked or coupled capacitors. Also, there are signal chains that amplify the CCD output and convert it from analog to digital. The Actel FPGA provides the timing, logic, and control functions in the CCD signal chain and inserts a camera ID into camera telemetry to simplify processing [399]. Figure 9.1 shows the Mars Exploration Rover.

Selected electronic parts have to undergo a multi-step flight consideration process before utilized in any space exploration mission [398, 400]. The first step is the general flight approval, during which the manufacturers perform additional space-grade verification tests beyond the normal commercial evaluation, and NASA meticulously examines the results. Additional device parameters, such as temperature considerations and semiconductor characteristics are verified in these tests. What follows is flight-specific approval. In this step, NASA engineers examine the device compatibility with the mission. For instance, considerations of the operating environment including factors like temperature and radiation. Also included are a variety of mission-specific situations that the robot may encounter and the associated risk assessment. Depending on the specific application of the device, whether mission-critical or not, and the expected mission lifetime, the risk standards vary. Finally, parts go through specific design considerations to ensure all the design requirements have been met. Parts are examined for their designs addressing issues such as SEL, SEU, and SEFI. The Xilinx FPGAs used addressed some of the SEE through the following methods [399]:

1. fabrication processes largely prevent SEL,

2. TMR reduces SEU frequency, and

3. scrubbing allows device recovery from single event functional interrupts.

MER went successful and despite being designed for only 90 Martian days (1 Martian day = 24.6 h), continued until 2019. The implementation of mitigation techniques was also proven to be effective as the observed error rate was very similar to that predicted [398].

9.3.2   MARS SCIENCE LABORATORY MISSION

As shown in Fig. 9.2, launched in 2011, Mars Science Lab (MSL) was the new Rover sent on to Mars. FPGAs were heavily used in their key components, mainly responsible for scientific instrument control, image processing, and communications.

Curiosity has 17 cameras on board: 4 navigation cameras, 8 hazard cameras, the Mars Hand Lens Imager (MAHLI), 2 Mast Cameras, the Mars Descent Imager (MARDI), and the ChemCam Remote Microscopic Imager [401]. MAHLI, the mast cameras, and MARDI share the same electronics design. Similar to the system used on MER, an Actel FPGA provides the timing, logic, and control functions in the CCD signal chain and transmits pixels to the digital electronics assembly (DEA), which interfaces the camera heads with the rover electronics, transmitting commands to the camera heads and data back to the rover. There is one DEA dedicated to each of the images above. Each has a Virtex-II FPGA that contains a Microblaze soft-processor core. All of the core functionalities of the DEA, including timing, interface, and compression, are implemented in the FPGAs as logic peripherals of the Microblaze. Specifically, the DEA provides an image processing pipeline that includes 12- to 8-bit commanding of input pixels, horizontal sub-framing, and lossless or JPEG image compression [401]. What

Figure 9.2: Mars science laboratory.

runs on the Microblaze is the DEA flight software, which coordinates DEA hardware functions such as camera movements. It receives and executes commands, and transmits commands from the Earth. The flight software also implements image acquisition algorithms, including auto-focus and auto-exposure, performs error correction of flash memory, and mechanism control fault protection [401]. In total, the flight software consists of 10,000 lines of ANSI C code, all implemented on the FPGAs. Additionally, FPGAs power communication boxes (Electra-Lite) to provide critical communication to Earth from the rovers through a Mars relay network [402]. They are responsible for a variety of high-speed bulk signal processing.

9.3.3   MARS 2020 MISSION

As shown in Fig. 9.3, Perseverance is NASA's latest launched Mars rover. The presence of FP-GAs continued and increased. FPGAs were used in the autonomous driving system as a co-processor for algorithm acceleration for the first time in NASA's planetary rovers. Perseverance runs on the GESTALT (grid-based estimation of surface traversability applied to local terrain) AutoNav algorithm, which is the same as Curiosity [403]. Added was the FPGA-based acceler-ator, called Vision Compute Element (VCE). During landing, VCE serves to provide sufficient compute power for the Lander Vision System (LVS), which performs an intensive task of esti-mating the landing location in 10 s by fusing data from the designed landing location, IMUs, and landmark matches. After landing, the connection between VCE and LVS is severed. Instead, VCE is re-purposed for the GESTALT driving algorithm. The VCE has three cards plugged into a PCI backplane: a CPU card with BAE RAD750 processor, a Compute Element Power Conditioning Unit (CEPCU), and a Computer Vision Acceleration Card (CVAC). While the former two parts were inherited from the MLS mission, the CVAC is new. The CVAC has two

Figure 9.3: Mars 2020.

FPGAs. One is called the Vision Processor–a Xilinx Virtex 5QV that contains image processing modules for matching landmarks to estimate positions. The other is called the Housekeeping FPGA–a Microsemi RTAX 2000 antifuse FPGA that handles tasks such as synchronization with the spacecraft, power management, and Vision Processor configuration.

9.4   SUMMARY

Through more than two decades of use in space, FPGAs have demonstrated their reliability and applicability for space robotic missions. The properties of FPGAs make them good on-board processors, ones that have high reliability, adaptability, processing power, and energy efficiency: FPGAs have unrivaled adaptability and can even be reconfigured in run time; their capability for high degree parallel processing allows significant acceleration in executing many complex algorithms; hardware/software co-design methods make them potentially more energy-efficient. FPGAs may finally help us close the two-decade performance gap between commercial processors and space-grade ASICs when it comes to powering space exploration robots. As a direct result, the achievements that the world has made in fields such as deep learning and computer vision, which were often too computationally intense for space-grade processors to be used, may become applicable for robots in space in the near future. The implementation of those new technologies will be of great benefit for space robots, boosting their autonomy and capabilities and allowing us to explore farther and faster.

CHAPTER 10

Conclusion

The commercialization of autonomous robots is a thriving sector, and likely to be the next major compute demand driver, after PC, cloud computing, and mobile computing. After examining various compute substrates for robotic computing, we believe that FPGAs are currently the best compute substrate for robotic applications for several reasons: first, robotic algorithms are still evolving rapidly, and thus any ASIC-based accelerators will be months or even years behind the state-of-the-art algorithms; on the other hand, FPGAs can be dynamically updated as needed. Second, robotic workloads are highly diverse, thus it is difficult for any ASIC-based robotic computing accelerator to reach economies of scale in the near future; on the other hand, FP-GAs are a cost-effective and energy-efficient alternative before one type of accelerator reaches economies of scale. Third, compared to SoCs that have reached economies of scale, e.g., mobile SoCs, FPGAs deliver a significant performance advantage.

10.1   WHAT WE HAVE COVERED IN THIS BOOK

For a robot to be intelligent, it must first be able to sense the world, to understand the environment through perception, to keep track of its own positions through localization, to synthesize all information so as to make decisions through planning and control. In addition to these basic tasks, robots often cooperate with each other in order to extend their perception capability and to make more intelligent decisions. All these tasks have to be performed in real-time and often on platforms with stringent energy constraints. As we believe that FPGAs are currently the best compute substrate for robotic computing, we have provided detailed descriptions and discussions on robotic computing on FPGAs in this book.

Specifically, in Chapter 2, we have provided the background of FPGA technologies for readers without prior FPGA knowledge to grasp the basic understanding of what an FPGA is and how an FPGA works. We have also introduced partial reconfiguration, a technique that exploits the flexibility of FPGAs and one that is extremely useful for various robotic workloads to time-share an FPGA so as to minimize energy consumption and resource utilization. In addition, we have explored existing techniques that enable the ROS, an essential infrastructure for robotic computing, to run directly on FPGAs.

In Chapter 3, we have introduced FPGA-based neural network accelerator designs for robotic perception and have demonstrated that with software-hardware co-design, FPGAs can achieve more than 10 times better speed and energy efficiency than the state-of-the-art GPUs. Hence, we have verified that FPGAs are a promising candidate for neural network acceleration.

In Chapter 4, we discussed various stereo vision algorithms in robotic perception and their FPGA accelerator designs. We also demonstrated that with careful algorithm-hardware co-design, FPGAs can achieve two orders of magnitude of higher energy efficiency and performance than the state-of-the-art GPUs and CPUs.

In Chapter 5, we introduced a general-purpose localization framework that integrates key primitives in existing algorithms along with its implementation in FPGA. We also showed the FPGA-based localization framework retains high accuracy of individual algorithms, simplifies the software stack, and provides a desirable acceleration target.

In Chapter 6, we discussed the motion planning module, and have compared multiple FPGA and ASIC accelerator designs in motion planning in order to analyze the intrinsic design trade-offs. We demonstrated that with careful algorithm-hardware co-design, FPGAs can achieve three orders of magnitude speedups compared to CPUs and two orders of magnitude than GPUs with significantly lower power consumption. This demonstrates that FPGAs can be a promising candidate for accelerating motion planning kernels.

In Chapter 7, we explored how FPGAs can be utilized in multi-robot exploration tasks. Specifically, we presented an FPGA-based interruptible CNN accelerator and a deployment framework for multi-robot exploration. We expect cooperative robots will become more popular as they bring significant advantages over single-robot systems, and thus optimizing cooperative robotic workloads on FPGAs will become increasingly important.

In Chapter 8, we provided a retrospective summary of PerceptIn's efforts on developing on-vehicle computing systems for autonomous vehicles, especially how FPGAs are utilized to accelerate critical tasks in a full autonomous driving stack. For instance, localization is accelerated on an FPGA while depth estimation and object detection are accelerated by a GPU. This case study demonstrated that FPGAs are capable of playing a crucial role in autonomous driving, and exploiting accelerator-level parallelism while taking into account constraints arising in different contexts could significantly improve on-vehicle processing.

In Chapter 9, we introduced how FPGAs have been utilized in space robotic applications in the past two decades. The properties of FPGAs make them good on-board processors for space missions, ones that have high reliability, adaptability, processing power, and energy efficiency. We believe that FPGAs may eventually help us close the two-decade performance gap between commercial processors and space-grade ASICs when it comes to powering space exploration robots.

10.2   LOOKING FORWARD

Looking forward, we still have a long way to go for FPGAs to become a first-class citizen in robotic computing substrates, as robotic computing on FPGAs is mainly constrained by infrastructure support, available tools, as well as talent supply. On infrastructure support, ROS is the most widely used operating system for autonomous machines and robots, as ROS provides the essential operating system services, including hardware abstraction, low-level device control, implementation of commonly used functionality, message-passing between processes, and package management. Unfortunately, FPGAs are not officially supported by ROS. Although there exist limited efforts to bridge ROS and FPGAs, to enable more robotic workloads on FPGAs, ROS needs to treat FPGAs as a first-class citizen like CPUs and GPUs.

On available tools, programming FPGAs is extremely challenging for engineers with limited hardware backgrounds, such as robotic engineers. Thus, an FPGA design and programming tool for robotic engineers is imperative. One recent effort, HeteroCL takes an initial step on addressing challenge [404]. Specifically, HeteroCL is comprised of a Python-based domain-specific language (DSL) and a compilation flow, potentially allowing programmers to port their robotic workloads onto FPGAs. We, as robotic computing researchers, shall develop more tools such as HeteroCL and have these tools officially integrated into the FPGA development cycle.

Last but not least, on talent supply, the existing engineering education system is not designed to provide cross-disciplinary training, as reconfigurable computing is usually an advanced topic in the Electrical and Computer Engineering department, whereas robotics is typically an advanced topic in the Computer Science department. To educate more engineers with cross-disciplinary backgrounds in robotic computing, we need to enhance our engineering education system with classes and projects on these cross-disciplinary topics as indicated in [405].

By exposing readers to various robotic modules and the state-of-the-art approaches of enabling and optimizing these modules on FPGAs, the exact purpose of this book is to guide interested researchers to the most fruitful research directions of robotic computing on FPGAs, as well as to guide interested students to master the essential skills to move the field of robotic computing forward. We sincerely hope that you have enjoyed reading this book.

Bibliography

[1] A. Qiantori, A. B. Sutiono, H. Hariyanto, H. Suwa, and T. Ohta, An emergency medical communications system by low altitude platform at the early stages of a natural disaster in Indonesia, *Journal of Medical Systems*, 36(1):41–52, 2012. DOI: 10.1007/s10916-010-9444-9. 1

[2] A. Ryan and J. K. Hedrick, A mode-switching path planner for UAV-assisted search and rescue, *Proc. of the 44th IEEE Conference on Decision and Control*, pages 1471–1476, 2005. DOI: 10.1109/cdc.2005.1582366. 1

[3] N. Smolyanskiy, A. Kamenev, J. Smith, and S. Birchfield, Toward low-flying autonomous MAV trail navigation using deep neural networks for environmental awareness, *ArXiv Preprint ArXiv:1705.02550*, 2017. DOI: 10.1109/iros.2017.8206285. 1

[4] A. Giusti, J. Guzzi, D. C. Cireşan, F.-L. He, J. P. Rodríguez, F. Fontana, M. Faessler, C. Forster, J. Schmidhuber, G. Di Caro et al., A machine learning approach to visual perception of forest trails for mobile robots, *IEEE Robotics and Automation Letters*, 1(2):661–667, 2015. DOI: 10.1109/lra.2015.2509024. 1

[5] B. Li, S. Liu, J. Tang, J.-L. Gaudiot, L. Zhang, and Q. Kong, Autonomous last-mile delivery vehicles in complex traffic environments, *Computer*, 53, 2020. DOI: 10.1109/mc.2020.2970924. 1

[6] S. J. Kim, Y. Jeong, S. Park, K. Ryu, and G. Oh, A survey of drone use for entertainment and AVR (augmented and virtual reality), *Augmented Reality and Virtual Reality*, pages 339–352, Springer, 2018. DOI: 10.1007/978-3-319-64027-3_23. 1

[7] S. Jung, S. Cho, D. Lee, H. Lee, and D. H. Shim, A direct visual servoing-based framework for the 2016 IROS autonomous drone racing challenge, *Journal of Field Robotics*, 35(1):146–166, 2018. DOI: 10.1002/rob.21743. 1

[8] Fact sheet—the federal aviation administration (FAA) aerospace forecast fiscal years (FY) 2020–2040. https://www.faa.gov/news/fact_sheets/news_story.cfm?newsId=24756, 2020. 1

[9] S. Liu, L. Li, J. Tang, S. Wu, and J.-L. Gaudiot, Creating autonomous vehicle systems, *Synthesis Lectures on Computer Science*, 6(1):i–186, 2017. DOI: 10.2200/s00787ed1v01y201707csl009. 1, 135, 138

[10] S. Krishnan, Z. Wan, K. Bhardwaj, P. Whatmough, A. Faust, G.-Y. Wei, D. Brooks, and V. J. Reddi, The sky is not the limit: A visual performance model for cyber-physical co-design in autonomous machines, *IEEE Computer Architecture Letters*, 19(1):38–42, 2020. DOI: 10.1109/lca.2020.2981022. 1

[11] Z. Wan, B. Yu, T. Y. Li, J. Tang, Y. Zhu, Y. Wang, A. Raychowdhury, and S. Liu, A survey of FPGA-based robotic computing, *ArXiv Preprint ArXiv:2009.06034*, 2020. 1

[12] S. Krishnan, Z. Wan, K. Bharadwaj, P. Whatmough, A. Faust, S. Neuman, G.-Y. Wei, D. Brooks, and V. J. Reddi, Machine learning-based automated design space exploration for autonomous aerial robots, *ArXiv Preprint ArXiv:2102.02988*, 2021. 1

[13] S. Liu and J.-L. Gaudiot, Autonomous vehicles lite self-driving technologies should start small, go slow, *IEEE Spectrum*, 57(3):36–49, 2020. DOI: 10.1109/mspec.2020.9014458. 1, 133

[14] S. Liu, L. Liu, J. Tang, B. Yu, Y. Wang, and W. Shi, Edge computing for autonomous driving: Opportunities and challenges, *Proc. of the IEEE*, 107(8):1697–1716, 2019. DOI: 10.1109/jproc.2019.2915983. 1, 133

[15] S. Liu, B. Yu, Y. Liu, K. Zhang, Y. Qiao, T. Y. Li, J. Tang, and Y. Zhu, The matter of time—a general and efficient system for precise sensor synchronization in robotic computing, *ArXiv Preprint ArXiv:2103.16045*, 2021. 1

[16] S. Liu, J. Peng, and J.-L. Gaudiot, Computer, drive my car! *Computer*, 1:8, 2017. DOI: 10.1109/mc.2017.2. 1, 2, 133, 140

[17] K. Guo, S. Zeng, J. Yu, Y. Wang, and H. Yang, [DL] a survey of FPGA-based neural network inference accelerators, *ACM Transactions on Reconfigurable Technology and Systems (TRETS)*, 12(1):1–26, 2019. DOI: 10.1145/3289185. 1

[18] S. Liu, J. Tang, C. Wang, Q. Wang, and J.-L. Gaudiot, A unified cloud platform for autonomous driving, *Computer*, 50(12):42–49, 2017. DOI: 10.1109/mc.2017.4451224. 2

[19] S. Liu, J. Tang, Z. Zhang, and J.-L. Gaudiot, Computer architectures for autonomous driving, *Computer*, 50(8):18–25, 2017. DOI: 10.1109/mc.2017.3001256. 2

[20] B. Yu, W. Hu, L. Xu, J. Tang, S. Liu, and Y. Zhu, Building the computing system for autonomous micromobility vehicles: Design constraints and architectural optimizations, *53rd Annual IEEE/ACM International Symposium on Microarchitecture (MICRO)*, 2020. DOI: 10.1109/micro50266.2020.00089. 3, 23

[21] S. Liu, B. Yu, Y. Liu, K. Zhang, Y. Qiao, T. Y. Li, J. Tang, and Y. Zhu, Brief industry paper: The matter of time—a general and efficient system for precise sensor synchronization in robotic computing, *IEEE Real-Time and Embedded Technology and Applications Symposium*, 2021. 4

[22] N. Dalal and B. Triggs, Histograms of oriented gradients for human detection, *IEEE Computer Society Conference on Computer Vision and Pattern Recognition (CVPR'05)*, 1:886–893, 2005. DOI: 10.1109/cvpr.2005.177. 4

[23] Xuming He, R. S. Zemel, and M. A. Carreira-Perpinan, Multiscale conditional random fields for image labeling, *Proc. of the IEEE Computer Society Conference on Computer Vision and Pattern Recognition, CVPR*, 2:II, 2004. DOI: 10.1109/CVPR.2004.1315232. 4

[24] X. He, R. S. Zemel, and D. Ray, Learning and incorporating top-down cues in image segmentation, *Computer Vision—ECCV*, A. Leonardis, H. Bischof, and A. Pinz, Eds., Berlin, Heidelberg: Springer Berlin Heidelberg, pages 338–351, 2006. DOI: 10.1007/11744047. 4

[25] Y. Xiang, A. Alahi, and S. Savarese, Learning to track: Online multi-object tracking by decision making, *IEEE International Conference on Computer Vision (ICCV)*, pages 4705–4713, 2015. DOI: 10.1109/iccv.2015.534. 5

[26] R. Girshick, Fast R-CNN, *IEEE International Conference on Computer Vision (ICCV)*, 2015. http://dx.doi.org/10.1109/ICCV.2015.169 DOI: 10.1109/iccv.2015.169. 5

[27] S. Ren, K. He, R. B. Girshick, and J. Sun, Faster R-CNN: Towards real-time object detection with region proposal networks, *CoRR*, 2015. http://arxiv.org/abs/1506.01497 DOI: 10.1109/tpami.2016.2577031. 5

[28] W. Liu, D. Anguelov, D. Erhan, C. Szegedy, S. E. Reed, C. Fu, and A. C. Berg, SSD: Single shot multibox detector, *CoRR*, 2015. http://arxiv.org/abs/1512.02325 DOI: 10.1007/978-3-319-46448-0_2. 5

[29] J. Redmon, S. K. Divvala, R. B. Girshick, and A. Farhadi, You only look once: Unified, real-time object detection, *CoRR*, 2015. http://arxiv.org/abs/1506.02640 DOI: 10.1109/cvpr.2016.91. 5

[30] J. Redmon and A. Farhadi, YOLO9000: Better, faster, stronger, *CoRR*, 2016. http://arxiv.org/abs/1612.08242 DOI: 10.1109/cvpr.2017.690. 5

[31] J. Long, E. Shelhamer, and T. Darrell, Fully convolutional networks for semantic segmentation, *CoRR*, 2014. http://arxiv.org/abs/1411.4038 DOI: 10.1109/cvpr.2015.7298965. 5

[32] K. He, X. Zhang, S. Ren, and J. Sun, Spatial pyramid pooling in deep convolutional networks for visual recognition, *CoRR*, 2014. http://arxiv.org/abs/1406.4729 DOI: 10.1007/978-3-319-10578-9_23. 5

[33] H. Zhao, J. Shi, X. Qi, X. Wang, and J. Jia, Pyramid scene parsing network, *CoRR*, 2016. http://arxiv.org/abs/1612.01105 DOI: 10.1109/cvpr.2017.660. 5

[34] L. Bertinetto, J. Valmadre, J. F. Henriques, A. Vedaldi, and P. H. S. Torr, Fully-convolutional siamese networks for object tracking, *CoRR*, 2016. http://arxiv.org/abs/1606.09549 DOI: 10.1007/978-3-319-48881-3_56. 5

[35] H. Durrant-Whyte and T. Bailey, Simultaneous localization and mapping: Part I, *IEEE Robotics Automation Magazine*, 13(2):99–110, 2006. DOI: 10.1109/mra.2006.1638022. 6

[36] M. Montemerlo, J. Becker, S. Bhat, H. Dahlkamp, D. Dolgov, S. Ettinger, D. Haehnel, T. Hilden, G. Hoffmann, B. Huhnke et al., Junior: The Stanford entry in the urban challenge, *Journal of Field Robotics*, 25(9):569–597, 2008. 6

[37] J. Ziegler, P. Bender, M. Schreiber, H. Lategahn, T. Strauss, C. Stiller, T. Dang, U. Franke, N. Appenrodt, C. G. Keller, E. Kaus, R. G. Herrtwich, C. Rabe, D. Pfeiffer, F. Lindner, F. Stein, F. Erbs, M. Enzweiler, C. Knöppel, J. Hipp, M. Haueis, M. Trepte, C. Brenk, A. Tamke, M. Ghanaat, M. Braun, A. Joos, H. Fritz, H. Mock, M. Hein, and E. Zeeb, Making bertha drive—an autonomous journey on a historic route, *IEEE Intelligent Transportation Systems Magazine*, 6(2):8–20, 2014. DOI: 10.1109/mits.2014.2306552. 6

[38] C. Katrakazas, M. Quddus, W.-H. Chen, and L. Deka, Real-time motion planning methods for autonomous on-road driving: State-of-the-art and future research directions, *Transportation Research Part C: Emerging Technologies*, 60:416–442, 2015. DOI: 10.1016/j.trc.2015.09.011. 6

[39] B. Paden, M. Čáp, S. Z. Yong, D. Yershov, and E. Frazzoli, A survey of motion planning and control techniques for self-driving urban vehicles, *IEEE Transactions on Intelligent Vehicles*, 1(1):33–55, 2016. DOI: 10.1109/tiv.2016.2578706. 6

[40] Y. Deng, Y. Chen, Y. Zhang, and S. Mahadevan, Fuzzy Dijkstra algorithm for shortest path problem under uncertain environment, *Applied Soft Computing*, 12(3):1231–1237, 2012. DOI: 10.1016/j.asoc.2011.11.011. 6

[41] P. E. Hart, N. J. Nilsson, and B. Raphael, A formal basis for the heuristic determination of minimum cost paths, *IEEE Transactions on Systems Science and Cybernetics*, 4(2):100–107, 1968. DOI: 10.1109/tssc.1968.300136. 6

[42] S. M. LaValle and J. J. Kuffner Jr, Randomized kinodynamic planning, *The International Journal of Robotics Research*, 20(5):378–400, 2001. DOI: 10.1177/02783640122067453. 6, 92

[43] L. E. Kavraki, P. Svestka, J.-C. Latombe, and M. H. Overmars, Probabilistic roadmaps for path planning in high-dimensional configuration spaces, *IEEE Transactions on Robotics and Automation*, 12(4):566–580, 1996. DOI: 10.1109/70.508439. 6, 7, 92

[44] S. Shalev-Shwartz, N. Ben-Zrihem, A. Cohen, and A. Shashua, Long-term planning by short-term prediction, *ArXiv Preprint ArXiv:1602.01580*, 2016. 6

[45] M. Gómez, R. González, T. Martínez-Marín, D. Meziat, and S. Sánchez, Optimal motion planning by reinforcement learning in autonomous mobile vehicles, *Robotica*, 30(2):159, 2012. DOI: 10.1017/s0263574711000452. 6, 7

[46] S. Shalev-Shwartz, S. Shammah, and A. Shashua, Safe, multi-agent, reinforcement learning for autonomous driving, *ArXiv Preprint ArXiv:1610.03295*, 2016. 6, 7

[47] M. Bojarski, D. Del Testa, D. Dworakowski, B. Firner, B. Flepp, P. Goyal, L. D. Jackel, M. Monfort, U. Muller, J. Zhang et al., End to end learning for self-driving cars, *ArXiv Preprint ArXiv:1604.07316*, 2016. 6

[48] X. Geng, H. Liang, B. Yu, P. Zhao, L. He, and R. Huang, A scenario-adaptive driving behavior prediction approach to urban autonomous driving, *Applied Sciences*, 7(4):426, 2017. DOI: 10.3390/app7040426. 6, 7

[49] C. J. Watkins and P. Dayan, Q-learning, *Machine Learning*, 8(3–4):279–292, 1992. 7

[50] V. R. Konda and J. N. Tsitsiklis, Actor-critic algorithms, *Advances in Neural Information Processing Systems*, pages 1008–1014, 2000. 7

[51] S. Liu, L. Li, J. Tang, S. Wu, and J.-L. Gaudiot, Creating autonomous vehicle systems, *Synthesis Lectures on Computer Science*, 8(2):i–216, 2020. DOI: 10.2200/s01036ed1v01y202007csl012. 7

[52] M. Berg, FPGA mitigation strategies for critical applications, 2019. https://ntrs.nasa.gov/citations/20190033455 12, 149, 150

[53] D. Sheldon, Flash-based FPGA NEPP FY12 summary report, 2013. https://nepp.nasa.gov/files/24179/12_101_JPL_Sheldon_Flash%20FPGA%20Summary%20Report_Sheldon%20rec%203_4_13.pdf 12

[54] I. Kuon, R. Tessier, and J. Rose, *FPGA Architecture: Survey and Challenges*, Now Publishers Inc., 2008. DOI: 10.1561/9781601981271. 12

[55] L. Liu, J. Tang, S. Liu, B. Yu, J.-L. Gaudiot, and Y. Xie, π-RT: A runtime framework to enable energy-efficient real-time robotic vision applications on heterogeneous architectures, *Computer*, 54, 2021. DOI: 10.1109/mc.2020.3015950. 14

[56] K. Vipin and S. A. Fahmy, FPGA dynamic and partial reconfiguration: A survey of architectures, methods, and applications, *ACM Computing Surveys (CSUR)*, 51(4):1–39, 2018. DOI: 10.1145/3193827. 16

[57] R. N. Pittman, Partial reconfiguration: A simple tutorial, *Technical Report*, 2012. 19

[58] S. Liu, R. N. Pittman, and A. Forin, Minimizing partial reconfiguration overhead with fully streaming DMA engines and intelligent ICAP controller, *FPGA*, Citeseer, page 292, 2010. DOI: 10.1145/1723112.1723190. 19, 20, 23

[59] J. H. Anderson and F. N. Najm, Active leakage power optimization for FPGAs, *IEEE Transactions on Computer-Aided Design of Integrated Circuits and Systems*, 25(3):423–437, 2006. DOI: 10.1109/tcad.2005.853692. 22

[60] S. Liu, R. N. Pittman, A. Forin, and J.-L. Gaudiot, Achieving energy efficiency through runtime partial reconfiguration on reconfigurable systems, *ACM Transactions on Embedded Computing Systems (TECS)*, 12(3):72, 2013. DOI: 10.1145/2442116.2442122. 22, 23

[61] E. Rublee, V. Rabaud, K. Konolige, and G. R. Bradski, ORB: An efficient alternative to sift or surf, *ICCV*, 11(1):2, Citeseer, 2011. DOI: 10.1109/iccv.2011.6126544. 23, 79

[62] B. D. Lucas and T. Kanade, An iterative image registration technique with an application to stereo vision, *Proc. of the 7th International Joint Conference on Artificial Intelligence*, 1981. 23, 79

[63] M. Quigley, K. Conley, B. Gerkey, J. Faust, T. Foote, J. Leibs, R. Wheeler, and A. Y. Ng, ROS: An open-source robot operating system, *ICRA Workshop on Open Source Software*, 3(3.2):5, Kobe, Japan, 2009. 24, 112, 113

[64] T. Ohkawa, K. Yamashina, T. Matsumoto, K. Ootsu, and T. Yokota, Architecture exploration of intelligent robot system using ROS-compliant FPGA component, *International Symposium on Rapid System Prototyping (RSP)*, *IEEE*, pages 1–7, 2016. DOI: 10.1145/2990299.2990312. 26

[65] K. Yamashina, T. Ohkawa, K. Ootsu, and T. Yokota, Proposal of ROS-compliant FPGA component for low-power robotic systems, *ArXiv Preprint ArXiv:1508.07123*, 2015. 26, 27

[66] Y. Sugata, T. Ohkawa, K. Ootsu, and T. Yokota, Acceleration of publish/subscribe messaging in ROS-compliant FPGA component, *Proc. of the 8th International Symposium on Highly Efficient Accelerators and Reconfigurable Technologies*, pages 1–6, 2017. DOI: 10.1145/3120895.3120904. 27

[67] L. Chappell, *Wireshark Network Analysis*, Podbooks.com, LLC, 2012. 28

[68] P. Merrick, S. Allen, and J. Lapp, XML remote procedure call (XML-RPC), U.S. Patent 7,028,312, 2006. 28

[69] H. Takase, T. Mori, K. Takagi, and N. Takagi, mROS: A lightweight runtime environment for robot software components onto embedded devices, *Proc. of the 10th International Symposium on Highly-Efficient Accelerators and Reconfigurable Technologies*, pages 1–6, 2019. DOI: 10.1145/3337801.3337815. 28

[70] H. Zhan, R. Garg, C. S. Weerasekera, K. Li, H. Agarwal, and I. Reid, Unsupervised learning of monocular depth estimation and visual odometry with deep feature reconstruction, *CVPR*, 2018. DOI: 10.1109/cvpr.2018.00043. 31

[71] R. Liu, J. Yang, Y. Chen, and W. Zhao, ESLAM: An energy-efficient accelerator for real-time ORB-SLAM on FPGA platform, *Proc. of the 56th Annual Design Automation Conference*, pages 1–6, 2019. DOI: 10.1145/3316781.3317820. 31, 73, 79, 89

[72] F. Radenović, G. Tolias, and O. Chum, Fine-tuning CNN image retrieval with no human annotation, *IEEE Transactions on Pattern Analysis and Machine Intelligence*, 41(7):1655–1668, 2018. DOI: 10.1109/tpami.2018.2846566. 31, 110, 111, 115, 128

[73] H. Jégou and A. Zisserman, Triangulation embedding and democratic aggregation for image search, *CVPR*, pages 3310–3317, 2014. DOI: 10.1109/cvpr.2014.417. 31, 110

[74] A. Krizhevsky, I. Sutskever, and G. E. Hinton, ImageNet classification with deep convolutional neural networks, *Advances in Neural Information Processing Systems*, pages 1097–1105, 2012. DOI: 10.1145/3065386. 32, 34, 37, 40

[75] O. Russakovsky, J. Deng, H. Su, J. Krause, S. Satheesh, S. Ma, Z. Huang, A. Karpathy, A. Khosla, M. Bernstein, A. C. Berg, and L. Fei-Fei, ImageNet large scale visual recognition challenge, *International Journal of Computer Vision (IJCV)*, 115(3):211–252, 2015. DOI: 10.1007/s11263-015-0816-y. 32

[76] R. Girshick, J. Donahue, T. Darrell, and J. Malik, Rich feature hierarchies for accurate object detection and semantic segmentation, *Proc. of the IEEE Conference on Computer Vision and Pattern Recognition*, pages 580–587, 2014. DOI: 10.1109/cvpr.2014.81. 32

[77] A. Hannun, C. Case, J. Casper, B. Catanzaro, G. Diamos, E. Elsen, R. Prenger, S. Satheesh, S. Sengupta, A. Coates et al., Deep speech: Scaling up end-to-end speech recognition, *ArXiv Preprint ArXiv:1412.5567*, 2014. 32, 34

[78] K. Simonyan and A. Zisserman, Very deep convolutional networks for large-scale image recognition, *ArXiv Preprint ArXiv:1409.1556*, 2014. 32, 34, 45

[79] A. G. Howard, M. Zhu, B. Chen, D. Kalenichenko, W. Wang, T. Weyand, M. Andreetto, and H. Adam, MobileNets: Efficient convolutional neural networks for mobile vision applications, *ArXiv Preprint ArXiv:1704.04861*, 2017. 32, 37

[80] X. Zhang, X. Zhou, M. Lin, and J. Sun, ShuffleNet: An extremely efficient convolutional neural network for mobile devices, *CoRR*, 2017. http://arxiv.org/abs/1707.01083 DOI: 10.1109/cvpr.2018.00716. 32

[81] Y. Jia, E. Shelhamer, J. Donahue, S. Karayev, J. Long, R. Girshick, S. Guadarrama, and T. Darrell, Caffe: Convolutional architecture for fast feature embedding, *Proc. of the 22nd ACM International Conference on Multimedia*, pages 675–678, 2014. DOI: 10.1145/2647868.2654889. 33, 117

[82] M. Abadi, A. Agarwal, P. Barham, E. Brevdo, Z. Chen, C. Citro, G. S. Corrado, A. Davis, J. Dean, M. Devin et al., TensorFlow: Large-scale machine learning on heterogeneous distributed systems, *ArXiv Preprint ArXiv:1603.04467*, 2016. 33

[83] C. Szegedy, W. Liu, Y. Jia, P. Sermanet, S. Reed, D. Anguelov, D. Erhan, V. Vanhoucke, and A. Rabinovich, Going deeper with convolutions, *Proc. of the IEEE Conference on Computer Vision and Pattern Recognition*, pages 1–9, 2015. DOI: 10.1109/cvpr.2015.7298594. 34

[84] K. He, X. Zhang, S. Ren, and J. Sun, Deep residual learning for image recognition, *Proc. of the IEEE Conference on Computer Vision and Pattern Recognition*, pages 770–778, 2016. DOI: 10.1109/cvpr.2016.90. 32, 34, 37, 45

[85] D. Amodei, S. Ananthanarayanan, R. Anubhai, J. Bai, E. Battenberg, C. Case, J. Casper, B. Catanzaro, Q. Cheng, G. Chen et al., Deep speech 2: End-to-end speech recognition in English and mandarin, *International Conference on Machine Learning*, pages 173–182, 2016. 34

[86] F. N. Iandola, S. Han, M. W. Moskewicz, K. Ashraf, W. J. Dally, and K. Keutzer, SqueezeNet: AlexNet-level accuracy with 50x fewer parameters and < 0.5 MB model size, *ArXiv Preprint ArXiv:1602.07360*, 2016. 37

[87] M. Tan, B. Chen, R. Pang, V. Vasudevan, and Q. V. Le, MnasNet: Platform-aware neural architecture search for mobile, *ArXiv Preprint ArXiv:1807.11626*, 2018. DOI: 10.1109/cvpr.2019.00293. 37

[88] X. Wang, F. Yu, Z.-Y. Dou, and J. E. Gonzalez, SkipNet: Learning dynamic routing in convolutional networks, *ArXiv Preprint ArXiv:1711.09485*, 2017. DOI: 10.1007/978-3-030-01261-8_25. 37

[89] H. Guan, S. Liu, X. Ma, W. Niu, B. Ren, X. Shen, Y. Wang, and P. Zhao, Cocopie: Enabling real-time AI on off-the-shelf mobile devices via compression-compilation co-design, *Communications of the ACM*, 2021. 37

[90] J. Qiu, J. Wang, S. Yao, K. Guo, B. Li, E. Zhou, J. Yu, T. Tang, N. Xu, S. Song et al., Going deeper with embedded FPGA platform for convolutional neural network, *Proc. of the ACM/SIGDA International Symposium on Field-Programmable Gate Arrays*, pages 26–35, 2016. DOI: 10.1145/2847263.2847265. 38, 39, 40, 41, 43, 46, 48, 49, 50, 111, 115, 117, 119, 121

[91] J. Wang, Q. Lou, X. Zhang, C. Zhu, Y. Lin, and D. Chen, Design flow of accelerating hybrid extremely low bit-width neural network in embedded FPGA, *ArXiv Preprint ArXiv:1808.04311*, 2018. DOI: 10.1109/fpl.2018.00035. 38

[92] K. Guo, L. Sui, J. Qiu, J. Yu, J. Wang, S. Yao, S. Han, Y. Wang, and H. Yang, Angel-eye: A complete design flow for mapping CNN onto embedded FPGA, *IEEE Transactions on Computer-Aided Design of Integrated Circuits and Systems*, 37(1):35–47, 2017. DOI: 10.1109/tcad.2017.2705069. 38, 39, 40, 43, 48, 49, 50, 111, 115, 117, 119, 121, 128

[93] T. Tambe, E.-Y. Yang, Z. Wan, Y. Deng, V. J. Reddi, A. Rush, D. Brooks, and G.-Y. Wei, AdaptivFloat: A floating-point based data type for resilient deep learning inference, *ArXiv Preprint ArXiv:1909.13271*, 2019. 38

[94] T. Tambe, E. Yang, Z. Wan, Y. Deng, V. J. Reddi, A. Rush, D. Brooks, and G.-Y. Wei, Algorithm-hardware co-design of adaptive floating-point encodings for resilient deep learning inference, *57th ACM/IEEE Design Automation Conference (DAC)*, pages 1–6, 2020. DOI: 10.1109/dac18072.2020.9218516. 38

[95] S. Krishnan, S. Chitlangia, M. Lam, Z. Wan, A. Faust, and V. J. Reddi, Quantized reinforcement learning (QUARL), *ArXiv Preprint ArXiv:1910.01055*, 2019. 38

[96] F. Li, B. Zhang, and B. Liu, Ternary weight networks, *ArXiv Preprint ArXiv:1605.04711*, 2016. 38, 39

[97] S. Zhou, Y. Wu, Z. Ni, X. Zhou, H. Wen, and Y. Zou, Dorefa-Net: Training low bitwidth convolutional neural networks with low bitwidth gradients, *ArXiv Preprint ArXiv:1606.06160*, 2016. 38, 39

[98] W. Chen, J. Wilson, S. Tyree, K. Weinberger, and Y. Chen, Compressing neural networks with the hashing trick, *International Conference on Machine Learning*, pages 2285–2294, 2015. 38

[99] S. Han, H. Mao, and W. J. Dally, Deep compression: Compressing deep neural networks with pruning, trained quantization and Huffman coding, *ArXiv Preprint ArXiv:1510.00149*, 2015. 38, 39, 40

[100] C. Zhu, S. Han, H. Mao, and W. J. Dally, Trained ternary quantization, *ArXiv Preprint ArXiv:1612.01064*, 2016. 38, 39

[101] X. Zhang, J. Zou, X. Ming, K. He, and J. Sun, Efficient and accurate approximations of nonlinear convolutional networks, *Proc. of the IEEE Conference on Computer Vision and Pattern Recognition*, pages 1984–1992, 2015. DOI: 10.1109/cvpr.2015.7298809. 40

[102] B. Liu, M. Wang, H. Foroosh, M. Tappen, and M. Pensky, Sparse convolutional neural networks, *Proc. of the IEEE Conference on Computer Vision and Pattern Recognition*, pages 806–814, 2015. DOI: 10.1109/CVPR.2015.7298681. 40

[103] A. Podili, C. Zhang, and V. Prasanna, Fast and efficient implementation of convolutional neural networks on FPGA, *Application-Specific Systems, Architectures and Processors (ASAP), IEEE 28th International Conference on*, pages 11–18, 2017. DOI: 10.1109/asap.2017.7995253. 40, 50

[104] H. Li, X. Fan, L. Jiao, W. Cao, X. Zhou, and L. Wang, A high performance FPGA-based accelerator for large-scale convolutional neural networks, *26th International Conference on Field Programmable Logic and Applications (FPL), IEEE*, pages 1–9, 2016. DOI: 10.1109/fpl.2016.7577308. 40, 45, 50, 51

[105] Q. Xiao, Y. Liang, L. Lu, S. Yan, and Y.-W. Tai, Exploring heterogeneous algorithms for accelerating deep convolutional neural networks on FPGAs, *Proc. of the 54th Annual Design Automation Conference, ACM*, page 62, 2017. DOI: 10.1145/3061639.3062244. 40, 43, 50

[106] Y. Guan, H. Liang, N. Xu, W. Wang, S. Shi, X. Chen, G. Sun, W. Zhang, and J. Cong, FP-DNN: An automated framework for mapping deep neural networks onto FPGAs with RTL-HLS hybrid templates, *Field-Programmable Custom Computing Machines (FCCM), IEEE 25th Annual International Symposium on*, pages 152–159, 2017. DOI: 10.1109/fccm.2017.25. 40, 49, 50

[107] C. Zhang, Z. Fang, P. Zhou, P. Pan, and J. Cong, Caffeine: Towards uniformed representation and acceleration for deep convolutional neural networks, *Computer-Aided Design (ICCAD), IEEE/ACM International Conference on*, pages 1–8, 2016. DOI: 10.1145/2966986.2967011. 40, 43, 50, 51

[108] S. Han, J. Kang, H. Mao, Y. Hu, X. Li, Y. Li, D. Xie, H. Luo, S. Yao, Y. Wang, H. Yang, and W. J. Dally, ESE: Efficient speech recognition engine with sparse LSTM on FPGA. *FPGA*, pages 75–84, 2017. DOI: 10.1145/3020078.3021745. 40, 45, 46, 49, 50, 52

[109] A. Prost-Boucle, A. Bourge, F. Pétrot, H. Alemdar, N. Caldwell, and V. Leroy, Scalable high-performance architecture for convolutional ternary neural networks on FPGA, *Field Programmable Logic and Applications (FPL), 27th International Conference on, IEEE*, pages 1–7, 2017. DOI: 10.23919/fpl.2017.8056850. 41

[110] E. Nurvitadhi, J. Sim, D. Sheffield, A. Mishra, S. Krishnan, and D. Marr, Accelerating recurrent neural networks in analytics servers: Comparison of FPGA, CPU, GPU, and ASIC, *Field Programmable Logic and Applications (FPL), 26th International Conference on, IEEE*, pages 1–4, 2016. DOI: 10.1109/fpl.2016.7577314. 41

[111] Y. Li, Z. Liu, K. Xu, H. Yu, and F. Ren, A 7.663-TOPS 8.2-W energy-efficient FPGA accelerator for binary convolutional neural networks, *ArXiv Preprint ArXiv:1702.06392*, 2017. DOI: 10.1145/3020078.3021786. 41

[112] H. Nakahara, H. Yonekawa, H. Iwamoto, and M. Motomura, A batch normalization free binarized convolutional deep neural network on an FPGA, *Proc. of the ACM/SIGDA International Symposium on Field-Programmable Gate Arrays*, pages 290–290, 2017. DOI: 10.1145/3020078.3021782. 41

[113] R. Zhao, W. Song, W. Zhang, T. Xing, J.-H. Lin, M. Srivastava, R. Gupta, and Z. Zhang, Accelerating binarized convolutional neural networks with software-programmable FPGAs, *Proc. of the ACM/SIGDA International Symposium on Field-Programmable Gate Arrays*, pages 15–24, 2017. DOI: 10.1145/3020078.3021741. 41

[114] Y. Umuroglu, N. J. Fraser, G. Gambardella, M. Blott, P. Leong, M. Jahre, and K. Vissers, FINN: A framework for fast, scalable binarized neural network inference, *Proc. of the ACM/SIGDA International Symposium on Field-Programmable Gate Arrays*, pages 65–74, 2017. DOI: 10.1145/3020078.3021744. 41

[115] H. Nakahara, T. Fujii, and S. Sato, A fully connected layer elimination for a binarizec convolutional neural network on an FPGA, *Field Programmable Logic and Applications (FPL), 27th International Conference on, IEEE*, pages 1–4, 2017. DOI: 10.23919/fpl.2017.8056771. 41, 49, 50

[116] L. Jiao, C. Luo, W. Cao, X. Zhou, and L. Wang, Accelerating low bit-width convolutional neural networks with embedded FPGA, *Field Programmable Logic and Applications (FPL), 27th International Conference on, IEEE*, pages 1–4, 2017. DOI: 10.23919/fpl.2017.8056820. 41, 49, 50

[117] D. J. Moss, E. Nurvitadhi, J. Sim, A. Mishra, D. Marr, S. Subhaschandra, and P. H. Leong, High performance binary neural networks on the xeon+ FPGA™ platform, *Field Programmable Logic and Applications (FPL), 27th International Conference on IEEE*, pages 1–4, 2017. DOI: 10.23919/fpl.2017.8056823. 41, 49, 50

[118] L. Yang, Z. He, and D. Fan, A fully onchip binarized convolutional neural network FPGA implementation with accurate inference, *Proc. of the International Symposium on Low Power Electronics and Design, ACM*, page 50, 2018. DOI: 10.1145/3218603.3218615. 41, 45

[119] M. Ghasemzadeh, M. Samragh, and F. Koushanfar, Rebnet: Residual binarized neural network, *IEEE 26th Annual International Symposium on Field-Programmable Custom Computing Machines (FCCM)*, 2018. DOI: 10.1109/fccm.2018.00018. 41

[120] M. Samragh, M. Ghasemzadeh, and F. Koushanfar, Customizing neural networks for efficient FPGA implementation, *Field-Programmable Custom Computing Machines (FCCM), IEEE 25th Annual International Symposium on*, pages 85–92, 2017. DOI: 10.1109/fccm.2017.43. 41

[121] J. Guo, S. Yin, P. Ouyang, L. Liu, and S. Wei, Bit-width based resource partitioning for CNN acceleration on FPGA, *Field-Programmable Custom Computing Machines (FCCM), IEEE 25th Annual International Symposium on*, page 31, 2017 DOI: 10.1109/fccm.2017.13. 41

[122] D. Nguyen, D. Kim, and J. Lee, Double MAC: Doubling the performance of convolutional neural networks on modern FPGAs, *Design, Automation and Test in Europe Conference and Exhibition (DATE), IEEE*, pages 890–893, 2017. DOI: 10.23919/date.2017.7927113. 42

[123] C. Zhang and V. K. Prasanna, Frequency domain acceleration of convolutional neural networks on CPU-FPGA shared memory system, *FPGA*, pages 35–44, 2017. DOI: 10.1145/3020078.3021727. 42, 43, 45, 46, 50, 51

[124] C. Ding, S. Liao, Y. Wang, Z. Li, N. Liu, Y. Zhuo, C. Wang, X. Qian, Y. Bai, G. Yuan, X. Ma, Y. Zhang, J. Tang, Q. Qiu, X. Lin, and B. Yuan, CirCNN: Accelerating and compressing deep neural networks using block-circulant weight matrices, *Proc. of the 50th Annual IEEE/ACM International Symposium on Microarchitecture*, pages 395–408, 2017. DOI: 10.1145/3123939.3124552. 42

[125] S. Winograd, *Arithmetic Complexity of Computations*, SIAM, 33, 1980. DOI: 10.1137/1.9781611970364. 43

[126] L. Lu, Y. Liang, Q. Xiao, and S. Yan, Evaluating fast algorithms for convolutional neural networks on FPGAs, *Field-Programmable Custom Computing Machines (FCCM), IEEE 25th Annual International Symposium on*, pages 101–108, 2017. DOI: 10.1109/fccm.2017.64. 43, 45, 50, 51, 52

[127] C. Zhuge, X. Liu, X. Zhang, S. Gummadi, J. Xiong, and D. Chen, Face recognition with hybrid efficient convolution algorithms on FPGAs, *Proc. of the on Great Lakes Symposium on VLSI, ACM*, pages 123–128, 2018. DOI: 10.1145/3194554.3194597. 43

[128] Y. Ma, Y. Cao, S. Vrudhula, and J.-S. Seo, Optimizing loop operation and dataflow in FPGA acceleration of deep convolutional neural networks, *Proc. of the ACM/SIGDA International Symposium on Field-Programmable Gate Arrays*, pages 45–54, 2017. DOI: 10.1145/3020078.3021736. 43, 45, 46, 48, 50

[129] J. Zhang and J. Li, Improving the performance of openCL-based FPGA accelerator for convolutional neural network, *FPGA*, pages 25–34, 2017. DOI: 10.1145/3020078.3021698. 43, 50, 51

[130] E. Wu, X. Zhang, D. Berman, and I. Cho, A high-throughput reconfigurable processing array for neural networks, *Field Programmable Logic and Applications (FPL), 27th International Conference on, IEEE*, pages 1–4, 2017. DOI: 10.23919/fpl.2017.8056794. 43

[131] https://github.com/Xilinx/chaidnn, 2018. 43

[132] https://www.xilinx.com/support/documentation/white_papers/wp504-accel-dnns.pdf, 2018. 43

[133] C. Zhang, P. Li, G. Sun, Y. Guan, B. Xiao, and J. Cong, Optimizing FPGA-based accelerator design for deep convolutional neural networks, *Proc. of the ACM/SIGDA International Symposium on Field-Programmable Gate Arrays*, pages 161–170, 2015. DOI: 10.1145/2684746.2689060. 45, 46, 48, 50, 51

[134] M. Motamedi, P. Gysel, V. Akella, and S. Ghiasi, Design space exploration of FPGA-based deep convolutional neural networks, *Design Automation Conference (ASP-DAC), 21st Asia and South Pacific, IEEE*, pages 575–580, 2016. DOI: 10.1109/asp-dac.2016.7428073. 45, 48

[135] Z. Liu, Y. Dou, J. Jiang, and J. Xu, Automatic code generation of convolutional neural networks in FPGA implementation, *Field-Programmable Technology (FPT), International Conference on, IEEE*, pages 61–68, 2016. DOI: 10.1109/FPT.2016.7929190. 45, 50, 51

[136] X. Zhang, J. Wang, C. Zhu, Y. Lin, J. Xiong, W.-M. Hwu, and D. Chen, DNNBuilder: An automated tool for building high-performance DNN hardware accelerators for FP-GAs, *Proc. of the International Conference on Computer-Aided Design, ACM*, page 56, 2018. DOI: 10.1145/3240765.3240801. 45

[137] C. Zhang, D. Wu, J. Sun, G. Sun, G. Luo, and J. Cong, Energy-Efficient CNN implementation on a deeply pipelined FPGA cluster, *Proc. of the International Symposium on Low Power Electronics and Design, ACM*, pages 326–331, 2016. DOI: 10.1145/2934583.2934644. 45, 50, 52

[138] Y. Shen, M. Ferdman, and P. Milder, Overcoming resource underutilization in spatial CNN accelerators, *Field Programmable Logic and Applications (FPL), 26th International Conference on, IEEE*, pages 1–4, 2016. DOI: 10.1109/fpl.2016.7577315. 45

[139] X. Lin, S. Yin, F. Tu, L. Liu, X. Li, and S. Wei, LCP: A layer clusters paralleling mapping method for accelerating inception and residual networks on FPGA, *Proc. of the 55th Annual Design Automation Conference, ACM*, page 16, 2018. DOI: 10.1145/3195970.3196067. 45

[140] X. Wei, C. H. Yu, P. Zhang, Y. Chen, Y. Wang, H. Hu, Y. Liang, and J. Cong, Automated systolic array architecture synthesis for high throughput CNN inference on FPGAs, *Proc. of the 54th Annual Design Automation Conference*, pages 1–6, 2017. DOI: 10.1145/3061639.3062207. 46

[141] U. Aydonat, S. O'Connell, D. Capalija, A. C. Ling, and G. R. Chiu, An openCL (TM) deep learning accelerator on arria 10, *ArXiv Preprint ArXiv:1701.03534*, 2017. 46, 50

[142] M. Horowitz, Energy table for 45 nm process, Stanford VLSI Wiki. https://sites.google.com/site/seecproject 48

[143] Y. Shen, M. Ferdman, and P. Milder, Escher: A CNN accelerator with flexible buffering to minimize off-chip transfer, *Proc. of the 25th IEEE International Symposium on Field-Programmable Custom Computing Machines (FCCM'17), IEEE Computer Society*, Los Alamitos, CA, 2017. DOI: 10.1109/fccm.2017.47. 48

[144] M. Alwani, H. Chen, M. Ferdman, and P. Milder, Fused-layer CNN accelerators, *Microarchitecture (MICRO), 49th Annual IEEE/ACM International Symposium on, IEEE*, pages 1–12, 2016. DOI: 10.1109/micro.2016.7783725. 49

[145] J. Yu, Y. Hu, X. Ning, J. Qiu, K. Guo, Y. Wang, and H. Yang, Instruction driven cross-layer CNN accelerator with winograd transformation on FPGA, in *International Conference on Field Programmable Technology*, 2017, pages 227–230. DOI: 10.1109/fpt.2017.8280147. 49

[146] N. Suda, V. Chandra, G. Dasika, A. Mohanty, Y. Ma, S. Vrudhula, J.-S. Seo, and Y. Cao, Throughput-optimized openCL-based FPGA accelerator for large-scale convolutional neural networks, *Proc. of the ACM/SIGDA International Symposium on Field-Programmable Gate Arrays*, pages 16–25, 2016. DOI: 10.1145/2847263.2847276. 50

[147] S. I. Venieris and C.-S. Bouganis, fpgaConvNet: Automated mapping of convolutional neural networks on FPGAs, *Proc. of the ACM/SIGDA International Symposium on Field-Programmable Gate Arrays*, pages 291–292, 2017. DOI: 10.1145/3020078.3021791. 50

[148] J. Shen, Y. Huang, Z. Wang, Y. Qiao, M. Wen, and C. Zhang, Towards a uniform template-based architecture for accelerating 2D and 3D CNNs on FPGA, *ACM/SIGDA International Symposium*, pages 97–106, 2018. DOI: 10.1145/3174243.3174257. 50

[149] Y. Guan, Z. Yuan, G. Sun, and J. Cong, FPGA-based accelerator for long short-term memory recurrent neural networks, *Design Automation Conference (ASP-DAC), 22nd Asia and South Pacific, IEEE*, pages 629–634, 2017. DOI: 10.1109/aspdac.2017.7858394. 50

[150] H. Mao, S. Han, J. Pool, W. Li, X. Liu, Y. Wang, and W. J. Dally, Exploring the regularity of sparse structure in convolutional neural networks, *ArXiv Preprint ArXiv:1705.08922*, 2017. 52

[151] http://www.deephi.com/technology/dnndk 2018. 53

[152] N. Dalal and B. Triggs, Histograms of oriented gradients for human detection, *IEEE Computer Society Conference on Computer Vision and Pattern Recognition (CVPR'05)*, 1:886–893, 2005. DOI: 10.1109/cvpr.2005.177. 55, 56

[153] P. Felzenszwalb, D. McAllester, and D. Ramanan, A discriminatively trained, multiscale, deformable part model, *IEEE Conference on Computer Vision and Pattern Recognition*, pages 1–8, 2008. DOI: 10.1109/cvpr.2008.4587597. 56

[154] X. He, R. S. Zemel, and M. Á. Carreira-Perpiñán, Multiscale conditional random fields for image labeling, *Proc. of the IEEE Computer Society Conference on Computer Vision and Pattern Recognition, CVPR*, 2:II, 2004. DOI: 10.1109/cvpr.2004.1315232. 56, 57

[155] X. He, R. S. Zemel, and D. Ray, Learning and incorporating top-down cues in image segmentation, *European Conference on Computer Vision*, pages 338–351, Springer, 2006. DOI: 10.1007/11744023_27. 56

[156] P. Krähenbühl and V. Koltun, Efficient inference in fully connected CRFS with Gaussian edge potentials, *Advances in Neural Information Processing Systems*, 24:109–117, 2011. 56

[157] L. Ladicky, C. Russell, P. Kohli, and P. H. Torr, Graph cut based inference with co-occurrence statistics, *European Conference on Computer Vision*, pages 239–253, Springer, 2010. DOI: 10.1007/978-3-642-15555-0_18. 56

[158] B. K. Horn and B. G. Schunck, Determining optical flow, *Techniques and Applications of Image Understanding*, 281:pages 319–331, International Society for Optics and Photonics, 1981. DOI: 10.1117/12.965761. 57

[159] S. L. Hicks, I. Wilson, L. Muhammed, J. Worsfold, S. M. Downes, and C. Kennard, A depth-based head-mounted visual display to aid navigation in partially sighted individuals, *PloS One*, 8(7):e67695, 2013. DOI: 10.1371/journal.pone.0067695. 58

[160] T. Whelan, R. F. Salas-Moreno, B. Glocker, A. J. Davison, and S. Leutenegger, Elas-ticFusion: Real-time dense slam and light source estimation, *The International Journal of Robotics Research*, 35(14):1697–1716, 2016. DOI: 10.1177/0278364916669237. 58

[161] V. A. Prisacariu, O. Kähler, S. Golodetz, M. Sapienza, T. Cavallari, P. H. Torr, and D. W. Murray, InfiniTAM v3: A framework for large-scale 3D reconstruction with loop closure, *ArXiv Preprint ArXiv:1708.00783*, 2017. 58

[162] S. Golodetz, T. Cavallari, N. A. Lord, V. A. Prisacariu, D. W. Murray, and P. H. Torr, Collaborative large-scale dense 3D reconstruction with online inter-agent pose optimisa-tion, *IEEE Transactions on Visualization and Computer Graphics*, 24(11):2895–2905, 2018. DOI: 10.1109/tvcg.2018.2868533. 58

[163] M. Pérez-Patricio and A. Aguilar-González, FPGA implementation of an efficient similarity-based adaptive window algorithm for real-time stereo matching, *Journal of Real-Time Image Processing*, 16(2):271–287, 2019. DOI: 10.1007/s11554-015-0530-6. 58

[164] D.-W. Yang, L.-C. Chu, C.-W. Chen, J. Wang, and M.-D. Shieh, Depth-reliability-based stereo-matching algorithm and its VLSI architecture design, *IEEE Trans-actions on Circuits and Systems for Video Technology*, 25(6):1038–1050, 2014. DOI: 10.1109/tcsvt.2014.2361419. 58

[165] A. Aguilar-González and M. Arias-Estrada, An FPGA stereo matching processor based on the sum of hamming distances, *International Symposium on Applied Reconfigurable Computing*, pages 66–77, Springer, 2016. DOI: 10.1007/978-3-319-30481-6_6. 58

[166] M. Pérez-Patricio, A. Aguilar-González, M. Arias-Estrada, H.-R. Hernandez-de Leon, J.-L. Camas-Anzueto, and J. de Jesús Osuna-Coutiño, An FPGA stereo matching unit based on fuzzy logic, *Microprocessors and Microsystems*, 42:87–99, 2016. DOI: 10.1016/j.micpro.2015.10.011. 58

[167] G. Cocorullo, P. Corsonello, F. Frustaci, and S. Perri, An efficient hardware-oriented stereo matching algorithm, *Microprocessors and Microsystems*, 46:21–33, 2016. DOI: 10.1016/j.micpro.2016.09.010. 58

[168] P. M. Santos, J. C. Ferreira, and J. S. Matos, Scalable hardware architecture for disparity map computation and object location in real-time, *Journal of Real-Time Image Processing*, 11(3):473–485, 2016. DOI: 10.1007/s11554-013-0338-1. 58

[169] K. M. Ali, R. B. Atitallah, N. Fakhfakh, and J.-L. Dekeyser, Exploring HLS optimiza-tions for efficient stereo matching hardware implementation, *International Symposium on Applied Reconfigurable Computing*, pages 168–176, Springer, 2017. DOI: 10.1007/978-3-319-56258-2_15. 58

[170] B. McCullagh, Real-time disparity map computation using the cell broadband engine, *Journal of Real-Time Image Processing*, 7(2):87–93, 2012. DOI: 10.1007/s11554-010-0155-8. 58

[171] L. Li, X. Yu, S. Zhang, X. Zhao, and L. Zhang, 3D cost aggregation with multiple minimum spanning trees for stereo matching, *Applied Optics*, 56(12):3411–3420, 2017. DOI: 10.1364/ao.56.003411. 58

[172] D. Zha, X. Jin, and T. Xiang, A real-time global stereo-matching on FPGA, *Microprocessors and Microsystems*, 47:419–428, 2016. DOI: 10.1016/j.micpro.2016.08.005. 58, 62, 71

[173] L. Puglia, M. Vigliar, and G. Raiconi, Real-time low-power FPGA architecture for stereo vision, *IEEE Transactions on Circuits and Systems II: Express Briefs*, 64(11):1307–1311, 2017. DOI: 10.1109/tcsii.2017.2691675. 58, 62, 71

[174] A. Kjær-Nielsen, K. Pauwels, J. B. Jessen, M. Van Hulle, N. Krüger et al., A two-level real-time vision machine combining coarse-and fine-grained parallelism, *Journal of Real-Time Image Processing*, 5(4):291–304, 2010. DOI: 10.1007/s11554-010-0159-4. 58

[175] H. Hirschmuller, Stereo processing by semiglobal matching and mutual information, *IEEE Transactions on Pattern Analysis and Machine Intelligence*, 30(2):328–341, 2007. DOI: 10.1109/tpami.2007.1166. 58

[176] A. Drory, C. Haubold, S. Avidan, and F. A. Hamprecht, Semi-global matching: A principled derivation in terms of message passing, *German Conference on Pattern Recognition*, pages 43–53, Springer, 2014. DOI: 10.1007/978-3-319-11752-2_4. 58

[177] S. K. Gehrig, F. Eberli, and T. Meyer, A real-time low-power stereo vision engine using semi-global matching, *International Conference on Computer Vision Systems*, pages 134–143, Springer, 2009. DOI: 10.1007/978-3-642-04667-4_14. 58, 62

[178] S. Wong, S. Vassiliadis, and S. Cotofana, A sum of absolute differences implementation in FPGA hardware, *Proc 28th Euromicro Conference, IEEE*, pages 183–188, 2002. DOI: 10.1109/eurmic.2002.1046155. 59

[179] M. Hisham, S. N. Yaakob, R. A. Raof, A. A. Nazren, and N. W. Embedded, Template matching using sum of squared difference and normalized cross correlation, *IEEE Student Conference on Research and Development (SCOReD), IEEE*, pages 100–104, 2015. DOI: 10.1109/scored.2015.7449303. 59

[180] J.-C. Yoo and T. H. Han, Fast normalized cross-correlation, *Circuits, Systems and Signal Processing*, 28(6):819, 2009. DOI: 10.1007/s00034-009-9130-7. 59

[181] B. Froba and A. Ernst, Face detection with the modified census transform, *6th IEEE International Conference on Automatic Face and Gesture Recognition, Proceedings*, pages 91–96, 2004. DOI: 10.1109/afgr.2004.1301514. 59

[182] O. Veksler, Fast variable window for stereo correspondence using integral images, *IEEE Computer Society Conference on Computer Vision and Pattern Recognition, Proceedings*, 1:I, 2003. DOI: 10.1109/cvpr.2003.1211403. 59

[183] A. Hosni, M. Bleyer, M. Gelautz, and C. Rhemann, Local stereo matching using geodesic support weights, *16th IEEE International Conference on Image Processing (ICIP)*, pages 2093–2096, 2009. DOI: 10.1109/icip.2009.5414478. 59

[184] O. Stankiewicz, G. Lafruit, and M. Domański, Multiview video: Acquisition, processing, compression, and virtual view rendering, *Academic Press Library in Signal Processing*, 6:3–74, Elsevier, 2018. DOI: 10.1016/b978-0-12-811889-4.00001-4. 60

[185] S. Jin, J. Cho, X. Dai Pham, K. M. Lee, S.-K. Park, M. Kim, and J. W. Jeon, FPGA design and implementation of a real-time stereo vision system, *IEEE Transactions on Circuits and Systems for Video Technology*, 20(1):15–26, 2009. DOI: 10.1109/tcsvt.2009.2026831. 60, 70, 71

[186] L. Zhang, K. Zhang, T. S. Chang, G. Lafruit, G. K. Kuzmanov, and D. Verkest, Real-time high-definition stereo matching on FPGA, *Proc. of the 19th ACM/SIGDA International Symposium on Field Programmable Gate Arrays*, pages 55–64, 2011. DOI: 10.1145/1950413.1950428. 60, 70, 71

[187] D. Honegger, P. Greisen, L. Meier, P. Tanskanen, and M. Pollefeys, Real-time velocity estimation based on optical flow and disparity matching, *IEEE/RSJ International Conference on Intelligent Robots and Systems*, pages 5177–5182, 2012. DOI: 10.1109/iros.2012.6385530. 61, 71

[188] M. Jin and T. Maruyama, Fast and accurate stereo vision system on FPGA, *ACM Transactions on Reconfigurable Technology and Systems (TRETS)*, 7(1):1–24, 2014. DOI: 10.1145/2567659. 61, 70, 71

[189] M. Werner, B. Stabernack, and C. Riechert, Hardware implementation of a full HD real-time disparity estimation algorithm, *IEEE Transactions on Consumer Electronics*, 60(1):66–73, 2014. DOI: 10.1109/tce.2014.6780927. 61

[190] S. Mattoccia and M. Poggi, A passive RGBD sensor for accurate and real-time depth sensing self-contained into an FPGA, *Proc. of the 9th International Conference on Distributed Smart Cameras*, pages 146–151, 2015. DOI: 10.1145/2789116.2789148. 61, 62

[191] V. C. Sekhar, S. Bora, M. Das, P. K. Manchi, S. Josephine, and R. Paily, Design and implementation of blind assistance system using real time stereo vision algorithms, *29th International Conference on VLSI Design and 15th International Conference on Embedded Systems (VLSID), IEEE*, pages 421–426, 2016. DOI: 10.1109/vlsid.2016.11. 61

[192] S. Perri, F. Frustaci, F. Spagnolo, and P. Corsonello, Stereo vision architecture for heterogeneous systems-on-chip, *Journal of Real-Time Image Processing*, 17(2):393–415, 2020. DOI: 10.1007/s11554-018-0782-z. 61

[193] Q. Yang, P. Ji, D. Li, S. Yao, and M. Zhang, Fast stereo matching using adaptive guided filtering, *Image and Vision Computing*, 32(3):202–211, 2014. DOI: 10.1016/j.imavis.2014.01.001. 61

[194] S. Park and H. Jeong, Real-time stereo vision FPGA chip with low error rate, *International Conference on Multimedia and Ubiquitous Engineering (MUE'07), IEEE*, pages 751–756, 2007. DOI: 10.1109/mue.2007.180. 61, 71

[195] S. Sabihuddin, J. Islam, and W. J. MacLean, Dynamic programming approach to high frame-rate stereo correspondence: A pipelined architecture implemented on a field programmable gate array, *Canadian Conference on Electrical and Computer Engineering, IEEE*, page 001 461–001 466, 2008. DOI: 10.1109/ccece.2008.4564784. 61, 71

[196] M. Jin and T. Maruyama, A real-time stereo vision system using a tree-structured dynamic programming on FPGA, *Proc. of the ACM/SIGDA International Symposium on Field Programmable Gate Arrays*, pages 21–24, 2012. DOI: 10.1145/2145694.2145698. 62, 70, 71

[197] R. Kamasaka, Y. Shibata, and K. Oguri, An FPGA-oriented graph cut algorithm for accelerating stereo vision, *International Conference on ReConFigurable Computing and FPGAs (ReConFig), IEEE*, pages 1–6, 2018. DOI: 10.1109/reconfig.2018.8641737. 62

[198] W. Wang, J. Yan, N. Xu, Y. Wang, and F.-H. Hsu, Real-time high-quality stereo vision system in FPGA, *IEEE Transactions on Circuits and Systems for Video Technology*, 25(10):1696–1708, 2015. DOI: 10.1109/tcsvt.2015.2397196. 62, 64, 70, 71

[199] D. Honegger, H. Oleynikova, and M. Pollefeys, Real-time and low latency embedded computer vision hardware based on a combination of FPGA and mobile CPU, *IEEE/RSJ International Conference on Intelligent Robots and Systems*, pages 4930–4935, 2014. DOI: 10.1109/iros.2014.6943263. 62

[200] C. Banz, S. Hesselbarth, H. Flatt, H. Blume, and P. Pirsch, Real-time stereo vision system using semi-global matching disparity estimation: Architecture and

FPGA-implementation, *International Conference on Embedded Computer Systems: Architectures, Modeling and Simulation, IEEE*, pages 93–101, 2010. DOI: 10.1109/icsamos.2010.5642077. 62, 70, 71

[201] L. F. Cambuim, J. P. Barbosa, and E. N. Barros, Hardware module for low-resource and real-time stereo vision engine using semi-global matching approach, *Proc. of the 30th Symposium on Integrated Circuits and Systems Design: Chip on the Sands*, pages 53–58, 2017. DOI: 10.1145/3109984.3109992. 63, 71

[202] L. F. Cambuim, L. A. Oliveira, E. N. Barros, and A. P. Ferreira, An FPGA-based real-time occlusion robust stereo vision system using semi-global matching, *Journal of Real-Time Image Processing*, pages 1–22, 2019. DOI: 10.1007/s11554-019-00902-w. 63, 71

[203] O. Rahnama, T. Cavalleri, S. Golodetz, S. Walker, and P. Torr, R3SGM: Real-time raster-respecting semi-global matching for power-constrained systems, *International Conference on Field-Programmable Technology (FPT), IEEE*, pages 102–109, 2018. DOI: 10.1109/fpt.2018.00025. 63, 66, 71

[204] J. Zhao, T. Liang, L. Feng, W. Ding, S. Sinha, W. Zhang, and S. Shen, FP-stereo: Hardware-efficient stereo vision for embedded applications, *ArXiv Preprint ArXiv:2006.03250*, 2020. DOI: 10.1109/fpl50879.2020.00052. 63, 71

[205] D. Hernandez-Juarez, A. Chacón, A. Espinosa, D. Vázquez, J. C. Moure, and A. M. López, Embedded real-time stereo estimation via semi-global matching on the GPU, *Procedia Computer Science*, 80:143–153, 2016. DOI: 10.1016/j.procs.2016.05.305. 63

[206] A. Geiger, M. Roser, and R. Urtasun, Efficient large-scale stereo matching, *Asian Conference on Computer Vision*, pages 25–38, Springer, 2010. DOI: 10.1007/978-3-642-19315-6_3. 63

[207] O. Rahnama, D. Frost, O. Miksik, and P. H. Torr, Real-time dense stereo matching with ELAS on FPGA-accelerated embedded devices, *IEEE Robotics and Automation Letters*, 3(3):2008–2015, 2018. DOI: 10.1109/lra.2018.2800786. 66, 67, 71

[208] O. Rahnama, T. Cavallari, S. Golodetz, A. Tonioni, T. Joy, L. Di Stefano, S. Walker, and P. H. Torr, Real-time highly accurate dense depth on a power budget using an FPGA-CPU hybrid SoC, *IEEE Transactions on Circuits and Systems II: Express Briefs*, 66(5):773–777, 2019. DOI: 10.1109/tcsii.2019.2909169. 66, 71

[209] T. Gao, Z. Wan, Y. Zhang, B. Yu, Y. Zhang, S. Liu, and A. Raychowdhury, iELAS: An ELAS-based energy-efficient accelerator for real-time stereo matching on FPGA platform, *ArXiv Preprint ArXiv:2104.05112*, 2021. 66, 67

[210] D. Scharstein and R. Szeliski, A taxonomy and evaluation of dense two-frame stereo correspondence algorithms, *International Journal of Computer Vision*, 47(1–3):7–42, 2002. DOI: 10.1109/smbv.2001.988771. 68

[211] M. Menze and A. Geiger, Object scene flow for autonomous vehicles, *Conference on Computer Vision and Pattern Recognition (CVPR)*, 2015. 68

[212] H. Hirschmuller and D. Scharstein, Evaluation of cost functions for stereo matching, *IEEE Conference on Computer Vision and Pattern Recognition, IEEE*, pages 1–8, 2007. DOI: 10.1109/cvpr.2007.383248. 68

[213] R. A. Hamzah, H. Ibrahim, and A. H. A. Hassan, Stereo matching algorithm based on illumination control to improve the accuracy, *Image Analysis and Stereology*, 35(1):39–52, 2016. DOI: 10.5566/ias.1369. 68

[214] Y. Shan, Z. Wang, W. Wang, Y. Hao, Y. Wang, K. Tsoi, W. Luk, and H. Yang, FPGA-based memory efficient high resolution stereo vision system for video tolling, *International Conference on Field-Programmable Technology, IEEE*, pages 29–32, 2012. DOI: 10.1109/fpt.2012.6412106. 70

[215] Y. Shan, Y. Hao, W. Wang, Y. Wang, X. Chen, H. Yang, and W. Luk, Hardware acceleration for an accurate stereo vision system using mini-census adaptive support region, *ACM Transactions on Embedded Computing Systems (TECS)*, 13(4):1–24, 2014. DOI: 10.1145/2584659. 70

[216] A. Kelly, *Mobile Robotics: Mathematics, Models, and Methods*, Cambridge University Press, 2013. https://books.google.com/books?id=laxZAQAAQBAJ DOI: 10.1017/cbo9781139381284. 73, 75, 139, 140, 147

[217] G. Dudek and M. Jenkin, *Computational Principles of Mobile Robotics*, Cambridge University Press, 2010. DOI: 10.1017/cbo9780511780929. 73

[218] Z. Li, Y. Chen, L. Gong, L. Liu, D. Sylvester, D. Blaauw, and H.-S. Kim, An 879GOPS 243 mW 80fps VGA fully visual CNN-slam processor for wide-range autonomous exploration, *IEEE International Solid-State Circuits Conference (ISSCC), IEEE*, pages 134–136, 2019. DOI: 10.1109/isscc.2019.8662397. 73, 79, 89

[219] A. Suleiman, Z. Zhang, L. Carlone, S. Karaman, and V. Sze, Navion: A 2 mW fully integrated real-time visual-inertial odometry accelerator for autonomous navigation of nano drones, *IEEE Journal of Solid-State Circuits*, 54(4):1106–1119, 2019. DOI: 10.1109/jssc.2018.2886342. 73, 76, 79, 89

[220] Z. Zhang, A. A. Suleiman, L. Carlone, V. Sze, and S. Karaman, Visual-inertial odometry on chip: An algorithm-and-hardware co-design approach, *Robotics: Science and Systems Online Proceedings*, 2017. DOI: 10.15607/rss.2017.xiii.028. 73, 79, 89

[221] J.-S. Yoon, J.-H. Kim, H.-E. Kim, W.-Y. Lee, S.-H. Kim, K. Chung, J.-S. Park, and L.-S. Kim, A graphics and vision unified processor with 0.89 μw/fps pose estimation engine for augmented reality, *IEEE International Solid-State Circuits Conference (ISSCC), IEEE*, pages 336–337, 2010. DOI: 10.1109/ISSCC.2010.5433907. 73, 89

[222] T. Kos, I. Markezic, and J. Pokrajcic, Effects of multipath reception on GPS positioning performance, *Proc. ELMAR, IEEE*, pages 399–402, 2010. 74

[223] N. El-Sheimy, S. Nassar, and A. Noureldin, Wavelet de-noising for IMU alignment, *IEEE Aerospace and Electronic Systems Magazine*, 19(10):32–39, 2004. DOI: 10.1109/maes.2004.1365016. 74

[224] A. Budiyono, L. Chen, S. Wang, K. McDonald-Maier, and H. Hu, Towards autonomous localization and mapping of AUVs: A survey, *International Journal of Intelligent Unmanned Systems*, 2013. DOI: 10.1108/20496421311330047. 75

[225] D. Gálvez-López and J. D. Tardos, Bags of binary words for fast place recognition in image sequences, *IEEE Transactions on Robotics*, 28(5):1188–1197, 2012. DOI: 10.1109/tro.2012.2197158. 76, 80

[226] R. Mur-Artal and J. D. Tardós, Fast relocalisation and loop closing in key frame-based slam, *IEEE International Conference on Robotics and Automation (ICRA)*, pages 846–853, 2014. DOI: 10.1109/icra.2014.6906953. 76, 80

[227] Dbow2. https://github.com/dorian3d/DBoW2 76

[228] iROBOT brings visual mapping and navigation to the roomba 980. https://spectrum.ieee.org/automaton/robotics/home-robots/irobot-brings-visual-mapping-and-navigation-to-the-roomba-980 76

[229] S. J. Julier and J. K. Uhlmann, Unscented filtering and nonlinear estimation, *Proc. of the IEEE*, 92(3):401–422, 2004. DOI: 10.1109/jproc.2003.823141. 76, 80

[230] A. I. Mourikis and S. I. Roumeliotis, A multi-state constraint Kalman filter for vision-aided inertial navigation, *Proc. IEEE International Conference on Robotics and Automation*, pages 3565–3572, 2007. DOI: 10.1109/robot.2007.364024. 76, 80, 88

[231] M. Li and A. I. Mourikis, High-precision, consistent EKF-based visual-inertial odometry, *The International Journal of Robotics Research*, 32(6):690–711, 2013. DOI: 10.1177/0278364913481251. 76

[232] Z. Zhang, S. Liu, G. Tsai, H. Hu, C.-C. Chu, and F. Zheng, PIRVS: An advanced visual-inertial slam system with flexible sensor fusion and hardware co-design, *IEEE International Conference on Robotics and Automation (ICRA)*, pages 1–7, 2018. DOI: 10.1109/icra.2018.8460672. 76, 80, 89

[233] K. Sun, K. Mohta, B. Pfrommer, M. Watterson, S. Liu, Y. Mulgaonkar, C. J. Taylor, and V. Kumar, Robust stereo visual inertial odometry for fast autonomous flight, *IEEE Robotics and Automation Letters*, 3(2):965–972, 2018. DOI: 10.1109/lra.2018.2793349. 76

[234] MSCKF vio. https://github.com/KumarRobotics/msckf_vio 76

[235] Visual inertial fusion. http://rpg.ifi.uzh.ch/docs/teaching/2018/13_visual_inertial_fusion_advanced.pdf#page=33 76

[236] C. Chen, X. Lu, A. Markham, and N. Trigoni, IONet: Learning to cure the curse of drift in inertial odometry, *32nd AAAI Conference on Artificial Intelligence*, 2018. 76

[237] D. Dusha and L. Mejias, Error analysis and attitude observability of a monocular GPS/visual odometry integrated navigation filter, *The International Journal of Robotics Research*, 31(6):714–737, 2012. DOI: 10.1177/0278364911433777. 76

[238] Mark: the world's first 4k drone positioned by visual inertial odometry. https://www.provideocoalition.com/mark-the-worlds-first-4k-drone-positioned-by-visual-inertial-odometry/ 76

[239] R. Hartley and A. Zisserman, *Multiple View Geometry in Computer Vision*, Cambridge University Press, 2003. DOI: 10.1017/cbo9780511811685. 76

[240] SLAMcore. https://www.slamcore.com/ 76

[241] Highly efficient machine learning for hololens. https://www.microsoft.com/en-us/research/uploads/prod/2018/03/Andrew-Fitzgibbon-Fitting-Models-to-Data-Accuracy-Speed-Robustness.pdf 76

[242] T. Qin, P. Li, and S. Shen, VINS-Mono: A robust and versatile monocular visual-inertial state estimator, *IEEE Transactions on Robotics*, 34(4):1004–1020, 2018. DOI: 10.1109/tro.2018.2853729. 76

[243] VINS-fusion. https://github.com/HKUST-Aerial-Robotics/VINS-Fusion 76, 84

[244] A. Geiger, P. Lenz, and R. Urtasun, Are we ready for autonomous driving? the KITTI vision benchmark suite, *IEEE Conference on Computer Vision and Pattern Recognition*, pages 3354–3361, 2012. DOI: 10.1109/cvpr.2012.6248074. 77, 86

[245] W. Qadeer, R. Hameed, O. Shacham, P. Venkatesan, C. Kozyrakis, and M. A. Horowitz, Convolution engine: Balancing efficiency and flexibility in specialized computing, *Proc. of the 40th IEEE Annual International Symposium on Computer Architecture*, 2013. DOI: 10.1145/2735841. 79

[246] Y. Feng, P. Whatmough, and Y. Zhu, ASV: Accelerated stereo vision system, *Proc. of the 52nd Annual IEEE/ACM International Symposium on Microarchitecture*, pages 643–656, 2019. DOI: 10.1145/3352460.3358253. 79, 139, 148

[247] E. Rosten and T. Drummond, Machine learning for high-speed corner detection, *European Conference on Computer Vision*, pages 430–443, Springer, 2006. DOI: 10.1007/11744023_34. 79

[248] M. Jakubowski and G. Pastuszak, Block-based motion estimation algorithms—a survey, *Opto-Electronics Review*, 21(1):86–102, 2013. DOI: 10.2478/s11772-013-0071-0. 79

[249] M. Calonder, V. Lepetit, C. Strecha, and P. Fua, Brief: Binary robust independent elementary features, *European Conference on Computer Vision*, pages 778–792, Springer, 2010. DOI: 10.1007/978-3-642-15561-1_56. 79, 153

[250] J. J. Moré, The Levenberg–Marquardt algorithm: Implementation and theory, *Numerical Analysis*, pages 105–116, Springer, 1978. DOI: 10.1007/bfb0067700. 80

[251] Ceres users. http://ceres-solver.org/users.html 80

[252] Vertex-7 datasheet. https://www.xilinx.com/support/documentation/data_sheets/ds180_7Series_Overview.pdf 85

[253] Xilinx, Zynq-7000 All Programmable SoC, http://www.xilinx.com/products/silicon-devices/soc/zynq-7000/ 2012. 86

[254] Tx1 datasheet. http://images.nvidia.com/content/tegra/embedded-systems/pdf/JTX1-Module-Product-sheet.pdf 86

[255] M. Burri, J. Nikolic, P. Gohl, T. Schneider, J. Rehder, S. Omari, M. W. Achtelik, and R. Siegwart, The EuRoC micro aerial vehicle datasets, *The International Journal of Robotics Research*, 35(10):1157–1163, 2016. DOI: 10.1177/0278364915620033. 86

[256] R. Mur-Artal and J. D. Tardós, ORB-SLAM2: An open-source slam system for monocular, stereo, and RGB-D cameras, *IEEE Transactions on Robotics*, 33(5):1255–1262, 2017. DOI: 10.1109/tro.2017.2705103. 88

[257] J. Engel, T. Schöps, and D. Cremers, LSD-SLAM: Large-scale direct monocular SLAM, *European Conference on Computer Vision*, pages 834–849, Springer, 2014. DOI: 10.1007/978-3-319-10605-2_54. 88

[258] A. Pumarola, A. Vakhitov, A. Agudo, A. Sanfeliu, and F. Moreno-Noguer, PL-SLAM: Real-time monocular visual SLAM with points and lines, *IEEE International Conference on Robotics and Automation (ICRA)*, pages 4503–4508, 2017. DOI: 10.1109/icra.2017.7989522. 88

[259] L. Marchetti, G. Grisetti, and L. Iocchi, A comparative analysis of particle filter based localization methods, *Robot Soccer World Cup*, pages 442–449, Springer, 2006. DOI: 10.1007/978-3-540-74024-7_44. 88

[260] L. Jetto, S. Longhi, and G. Venturini, Development and experimental validation of an adaptive extended Kalman filter for the localization of mobile robots, *IEEE Transactions on Robotics and Automation*, 15(2):219–229, 1999. DOI: 10.1109/70.760343. 88

[261] G. Mao, S. Drake, and B. D. Anderson, Design of an extended Kalman filter for UAV localization, *Information, Decision and Control, IEEE*, pages 224–229, 2007. DOI: 10.1109/idc.2007.374554. 88

[262] R. Mur-Artal, J. M. M. Montiel, and J. D. Tardos, ORB-SLAM: A versatile and accurate monocular SLAM system, *IEEE Transactions on Robotics*, 31(5):1147–1163, 2015. DOI: 10.1109/tro.2015.2463671. 88

[263] W. Fang, Y. Zhang, B. Yu, and S. Liu, FPGA-based ORB feature extraction for real-time visual slam, *International Conference on Field Programmable Technology (ICFPT), IEEE*, pages 275–278, 2017. DOI: 10.1109/fpt.2017.8280159. 89, 134

[264] Q. Gautier, A. Althoff, and R. Kastner, FPGA architectures for real-time dense SLAM, *IEEE 30th International Conference on Application-Specific Systems, Architectures and Processors (ASAP)*, 2160:83–90, 2019. DOI: 10.1109/asap.2019.00-25. 89

[265] K. Boikos and C.-S. Bouganis, Semi-dense slam on an FPGA SoC, *26th International Conference on Field Programmable Logic and Applications (FPL), IEEE*, pages 1–4, 2016. DOI: 10.1109/fpl.2016.7577365. 89

[266] D. T. Tertei, J. Piat, and M. Devy, FPGA design of EKF block accelerator for 3D visual SLAM, *Computers and Electrical Engineering*, 55:123–137, 2016. DOI: 10.1016/j.compeleceng.2016.05.003. 89

[267] S. Karaman and E. Frazzoli, Sampling-based algorithms for optimal motion planning, *The International Journal of Robotics Research*, 30(7):846–894, 2011. DOI: 10.1177/0278364911406761. 91, 92, 94

[268] J. D. Gammell, S. S. Srinivasa, and T. D. Barfoot, Batch informed trees (bit*): Sampling-based optimal planning via the heuristically guided search of implicit random geometric graphs, *IEEE International Conference on Robotics and Automation (ICRA)*, pages 3067–3074, 2015. DOI: 10.1109/icra.2015.7139620. 91

[269] K. Hauser, Lazy collision checking in asymptotically-optimal motion planning, *IEEE International Conference on Robotics and Automation (ICRA)*, pages 2951–2957, 2015. DOI: 10.1109/icra.2015.7139603. 91

[270] J. Pan, C. Lauterbach, and D. Manocha, g-Planner: Real-time motion planning and global navigation using GPUs, *AAAI*, 2010. 91, 95, 96

[271] J. Pan and D. Manocha, GPU-based parallel collision detection for fast motion planning, *The International Journal of Robotics Research*, 31(2):187–200, 2012. DOI: 10.1177/0278364911429335. 91, 95, 96

[272] S. Murray, W. Floyd-Jones, Y. Qi, D. J. Sorin, and G. Konidaris, Robot motion planning on a chip, *Robotics: Science and Systems*, 2016. DOI: 10.15607/rss.2016.xii.004. 91, 97, 98, 99, 100, 103, 104, 105

[273] S. Murray, W. Floyd-Jones, G. Konidaris, and D. J. Sorin, A programmable architecture for robot motion planning acceleration, *IEEE 30th International Conference on Application-specific Systems, Architectures and Processors (ASAP)*, 2160:185–188, 2019. DOI: 10.1109/asap.2019.000-4. 91, 97, 99, 100, 103, 104, 105, 106

[274] J. H. Reif, Complexity of the mover's problem and generalizations, *20th Annual Symposium on Foundations of Computer Science (SFCS), IEEE*, pages 421–427, 1979. DOI: 10.1109/sfcs.1979.10. 92

[275] J. Barraquand, L. Kavraki, J.-C. Latombe, R. Motwani, T.-Y. Li, and P. Raghavan, A random sampling scheme for path planning, *The International Journal of Robotics Research*, 16(6):759–774, 1997. DOI: 10.1177/027836499701600604. 92

[276] R. Bohlin and L. E. Kavraki, Path planning using lazy PRM, *Proc. ICRA, Millennium Conference, IEEE International Conference on Robotics and Automation, Symposia Proceedings (Cat. no. 00CH37065)*, 1:521–528, 2000. DOI: 10.1109/robot.2000.844107. 93

[277] A. Short, Z. Pan, N. Larkin, and S. Van Duin, Recent progress on sampling based dynamic motion planning algorithms, *IEEE International Conference on Advanced Intelligent Mechatronics (AIM)*, pages 1305–1311, 2016. DOI: 10.1109/aim.2016.7576950. 94

[278] J. J. Kuffner and S. M. LaValle, RRT-connect: An efficient approach to single-query path planning, *Proc. ICRA, Millennium Conference, IEEE International Conference on Robotics and Automation, Symposia Proceedings (Cat. no. 00CH37065)*, 2:995–1001, 2000. DOI: 10.1109/robot.2000.844730. 94

[279] D. Hsu, J.-C. Latombe, and R. Motwani, Path planning in expansive configuration spaces, *Proc. of International Conference on Robotics and Automation, IEEE*, 3:2719–2726, 1997. DOI: 10.1109/robot.1997.619371. 94

[280] E. Plaku, K. E. Bekris, B. Y. Chen, A. M. Ladd, and L. E. Kavraki, Sampling-based roadmap of trees for parallel motion planning, *IEEE Transactions on Robotics*, 21(4):597–608, 2005. DOI: 10.1109/tro.2005.847599. 94

[281] J. Bialkowski, S. Karaman, and E. Frazzoli, Massively parallelizing the RRT and the RRT, *IEEE/RSJ International Conference on Intelligent Robots and Systems*, pages 3513–3518, 2011. DOI: 10.1109/IROS.2011.6095053. 94, 95

[282] N. Atay and B. Bayazit, A motion planning processor on reconfigurable hardware, *Proc. IEEE International Conference on Robotics and Automation, ICRA*, pages 125–132, 2006. DOI: 10.1109/robot.2006.1641172. 97

[283] S. Lian, Y. Han, X. Chen, Y. Wang, and H. Xiao, Dadu-P: A scalable accelerator for robot motion planning in a dynamic environment, *55th ACM/ESDA/IEEE Design Automation Conference (DAC)*, pages 1–6, 2018. DOI: 10.1109/dac.2018.8465785. 97, 100, 101, 103, 104, 105

[284] Y. Yang, X. Chen, and Y. Han, Dadu-CD: Fast and efficient processing-in-memory accelerator for collision detection, *57th ACM/IEEE Design Automation Conference (DAC)*, pages 1–6, 2020. DOI: 10.1109/dac18072.2020.9218709. 97, 102, 103, 104, 105

[285] S. Murray, W. Floyd-Jones, Y. Qi, G. Konidaris, and D. J. Sorin, The microarchitecture of a real-time robot motion planning accelerator, *49th Annual IEEE/ACM International Symposium on Microarchitecture (MICRO)*, pages 1–12, 2016. DOI: 10.1109/micro.2016.7783748. 98, 99, 100, 147, 148

[286] Y. Han, Y. Yang, X. Chen, and S. Lian, Dadu series-fast and efficient robot accelerators, *IEEE/ACM International Conference on Computer Aided Design (ICCAD)*, pages 1–8, 2020. DOI: 10.1145/3400302.3415759. 101

[287] S. Ghose, A. Boroumand, J. S. Kim, J. Gómez-Luna, and O. Mutlu, Processing-in-memory: A workload-driven perspective, *IBM Journal of Research and Development*, 63(6):3–1, 2019. DOI: 10.1147/jrd.2019.2934048. 102

[288] V. Sze, Y.-H. Chen, T.-J. Yang, and J. S. Emer, Efficient processing of deep neural networks, *Synthesis Lectures on Computer Architecture*, 15(2):1–341, 2020. DOI: 10.2200/s01004ed1v01y202004cac050. 102

[289] E. W. Dijkstra et al., A note on two problems in connexion with graphs, *Numerische Mathematik*, 1(1):269–271, 1959. DOI: 10.1007/bf01386390. 106

[290] R. Bellman, On a routing problem, *Quarterly of Applied Mathematics*, 16(1):87–90, 1958. DOI: 10.1090/qam/102435. 106

[291] R. W. Floyd, Algorithm 97: Shortest path, *Communications of the ACM*, 5(6):345, 1962. DOI: 10.1145/367766.368168. 106

[292] Y. Takei, M. Hariyama, and M. Kameyama, Evaluation of an FPGA-based shortest-path-search accelerator, *Proc. of the International Conference on Parallel and Distributed Processing Techniques and Applications (PDPTA). The Steering Committee of The World Congress in Computer Science, Computer Engineering and Applied Computing (WorldComp)*, page 613, 2015. 106, 107

[293] G. Lei, Y. Dou, R. Li, and F. Xia, An FPGA implementation for solving the large single-source-shortest-path problem, *IEEE Transactions on Circuits and Systems II: Express Briefs*, 63(5):473–477, 2015. DOI: 10.1109/tcsii.2015.2505998. 106, 108

[294] P. Harish and P. Narayanan, Accelerating large graph algorithms on the GPU using cuda, *International Conference on High-Performance Computing*, pages 197–208, Springer, 2007. DOI: 10.1007/978-3-540-77220-0_21. 106

[295] G. J. Katz and J. T. Kider, All-pairs shortest-paths for large graphs on the GPU, *Proc. of the 23rd ACM SIGGRAPH/Eurographics Symposium on Graphics Hardware*, 2008. 106

[296] G. Malewicz, M. H. Austern, A. J. Bik, J. C. Dehnert, I. Horn, N. Leiser, and G. Czajkowski, Pregel: A system for large-scale graph processing, *Proc. of the ACM SIG-MOD International Conference on Management of data*, pages 135–146, 2010. DOI: 10.1145/1807167.1807184. 106

[297] K. Sridharan, T. Priya, and P. R. Kumar, Hardware architecture for finding shortest paths, *TENCON IEEE Region 10 Conference*, pages 1–5, 2009. DOI: 10.1109/tencon.2009.5396155. 106

[298] S. Zhou, C. Chelmis, and V. K. Prasanna, Accelerating large-scale single-source shortest path on FPGA, *IEEE International Parallel and Distributed Processing Symposium Workshop*, pages 129–136, 2015. DOI: 10.1109/ipdpsw.2015.130. 107

[299] U. Bondhugula, A. Devulapalli, J. Dinan, J. Fernando, P. Wyckoff, E. Stahlberg, and P. Sadayappan, Hardware/software integration for FPGA-based all-pairs shortest-paths, *14th Annual IEEE Symposium on Field-Programmable Custom Computing Machines*, pages 152–164, 2006. DOI: 10.1109/fccm.2006.48. 107

[300] S. Hougardy, The Floyd–Warshall algorithm on graphs with negative cycles, *Information Processing Letters*, 110(8–9):279–281, 2010. DOI: 10.1016/j.ipl.2010.02.001. 107

[301] S. Liu, B. Yu, J. Tang, and Q. Zhu, Towards fully intelligent transportation through infrastructure-vehicle cooperative autonomous driving: Challenges and opportunities, *ArXiv Preprint ArXiv:2103.02176*, 2021. 109

[302] M. Corah, C. O'Meadhra, K. Goel, and N. Michael, Communication-efficient planning and mapping for multi-robot exploration in large environments, *IEEE Robotics and Automation Letters*, 4(2):1715–1721, 2019. DOI: 10.1109/lra.2019.2897368. 109

[303] H. G. Tanner and A. Kumar, Towards decentralization of multi-robot navigation functions, *ICRA, IEEE*, pages 4132–4137, 2005. DOI: 10.1109/robot.2005.1570754. 109

[304] J. L. Baxter, E. Burke, J. M. Garibaldi, and M. Norman, Multi-robot search and rescue: A potential field based approach, *Autonomous Robots and Agents*, pages 9–16, Springer, 2007. DOI: 10.1007/978-3-540-73424-6_2. 109

[305] T. Cieslewski, S. Choudhary, and D. Scaramuzza, Data-efficient decentralized visual SLAM, *ICRA, IEEE*, pages 2466–2473, 2018. DOI: 10.1109/icra.2018.8461155. 109, 128

[306] B.-J. Ho, P. Sodhi, P. Teixeira, M. Hsiao, T. Kusnur, and M. Kaess, Virtual occupancy grid map for submap-based pose graph slam and planning in 3D environments, *IEEE/RSJ International Conference on Intelligent Robots and Systems (IROS)*, pages 2175–2182, 2018. DOI: 10.1109/iros.2018.8594234. 109

[307] P. Sodhi, B.-J. Ho, and M. Kaess, Online and consistent occupancy grid mapping for planning in unknown environments, *IEEE/RSJ International Conference on Intelligent Robots and Systems (IROS)*, pages 7879–7886, 2019. DOI: 10.1109/iros40897.2019.8967991. 109

[308] S. Choudhary, L. Carlone, C. Nieto, J. Rogers, H. I. Christensen, and F. Dellaert, Distributed mapping with privacy and communication constraints: Lightweight algorithms and object-based models, *The International Journal of Robotics Research*, 36:1286–1311, 2017. DOI: 10.1177/0278364917732640. 109, 128

[309] C. Wu, S. Agarwal, B. Curless, and S. M. Seitz, Multicore bundle adjustment, *CVPR, IEEE*, pages 3057–3064, 2011. DOI: 10.1109/cvpr.2011.5995552. 110

[310] UltraScale MPSoC Architecture, 2019. https://www.xilinx.com/products/technology/ultrascale-mpsoc.html 110, 127

[311] D. DeTone, T. Malisiewicz, and A. Rabinovich, SuperPoint: Self-supervised interest point detection and description, *Proc. of the IEEE Conference on Computer Vision and Pattern Recognition Workshops*, pages 224–236, 2018. DOI: 10.1109/cvprw.2018.00060. 110, 111, 115, 128

[312] E. Simo-Serra, E. Trulls, L. Ferraz, I. Kokkinos, P. Fua, and F. Moreno-Noguer, Discriminative learning of deep convolutional feature point descriptors, *Proc. of the IEEE International Conference on Computer Vision*, pages 118–126, 2015. DOI: 10.1109/iccv.2015.22. 110

[313] K. M. Yi, E. Trulls, V. Lepetit, and P. Fua, Lift: Learned invariant feature transform, *ECCV*, pages 467–483, Springer, 2016. DOI: 10.1007/978-3-319-46466-4_28. 110

[314] R. Arandjelovic, P. Gronat, A. Torii, T. Pajdla, and J. Sivic, NetVLAD: CNN architecture for weakly supervised place recognition, *Proc. of the IEEE Conference on Computer Vision and Pattern Recognition*, pages 5297–5307, 2016. DOI: 10.1109/cvpr.2016.572. 110

[315] R. Mur-Artal and J. D. Tards, ORB-SLAM2: An open-source SLAM system for monocular, stereo, and RGB-D cameras, *IEEE Transactions on Robotics*, 33:1255–1262, 2016. DOI: 10.1109/tro.2017.2705103. 110

[316] J. Long, E. Shelhamer, and T. Darrell, Fully convolutional networks for semantic segmentation, *CVPR*, pages 3431–3440, 2015. DOI: 10.1109/cvpr.2015.7298965. 111

[317] S. Ren, K. He, R. Girshick, and J. Sun, Faster R-CNN: Towards real-time object detection with region proposal networks, *Advances in Neural Information Processing Systems*, pages 91–99, 2015. DOI: 10.1109/tpami.2016.2577031. 111

[318] J. Yu, G. Ge, Y. Hu, X. Ning, J. Qiu, K. Guo, Y. Wang, and H. Yang, Instruction driven cross-layer CNN accelerator for fast detection on FPGA, *ACM Transactions on Reconfigurable Technology and Systems (TRETS)*, 11(3):1–23, 2018. DOI: 10.1145/3283452. 111, 115, 117

[319] H. Li, X. Fan, L. Jiao, W. Cao, X. Zhou, and L. Wang, A high performance FPGA-based accelerator for large-scale convolutional neural networks, *FPL, IEEE*, pages 1–9, 2016. DOI: 10.1109/fpl.2016.7577308. 111

[320] L. Lu, Y. Liang, Q. Xiao, and S. Yan, Evaluating fast algorithms for convolutional neural networks on FPGAs, *FCCM*, pages 101–108, 2017. DOI: 10.1109/fccm.2017.64. 111

[321] Xilinx Zynq UltraScale+ MPSoC ZCU102 Evaluation Kit, 2019. https://www.xilinx.com/products/boards-and-kits/ek-u1-zcu102-g.html 111, 127

[322] M. Mohanan and A. Salgoankar, A survey of robotic motion planning in dynamic environments, *Robotics and Autonomous Systems*, 100:171–185, 2018. DOI: 10.1016/j.robot.2017.10.011. 112

[323] R. Ramsauer, J. Kiszka, D. Lohmann, and W. Mauerer, Look mum, no VM exits! (almost), *CoRR*. http://arxiv.org/abs/1705.06932 112

[324] D. Jen and A. Lotan, Processor interrupt system, January 29 1974, U.S. Patent 3,789,365. 112

[325] DNNDK User Guide—Xilinx, 2019. https://www.xilinx.com/support/documentation/user_guides/ug1327-dnndk-user-guide.pdf 115, 117

[326] Softmax function—Wikipedia, 2019. https://en.wikipedia.org/wiki/Softmax_function 115

[327] Norm (mathematics)—Wikipedia, 2019. https://en.wikipedia.org/wiki/Norm_(mathematics)#Euclidean_norm 115

[328] A. Neubeck and L. J. V. Gool, Efficient non-maximum suppression, *International Conference on Pattern Recognition*, 2006. DOI: 10.1109/icpr.2006.479. 115

[329] R. Banakar, S. Steinke, B. S. Lee, M. Balakrishnan, and P. Marwedel, Scratchpad memory: A design alternative for cache on-chip memory in embedded systems, *CODES*, 2002. DOI: 10.1109/codes.2002.1003604. 115

[330] S. B. Furber, *ARM System-on-Chip Architecture*, Pearson Education, 2000. 119

[331] S. Shah, D. Dey, C. Lovett, and A. Kapoor, AirSim: High-fidelity visual and physical simulation for autonomous vehicles, *Field and Service Robotics*, pages 621–635, Springer, 2018. DOI: 10.1007/978-3-319-67361-5_40. 127, 131

[332] V. Lepetit, F. Moreno-Noguer, and P. Fua, EPnP: An accurate O(n) solution to the PnP problem, *International Journal of Computer Vision*, 81(2):155–166, 2009. DOI: 10.1007/s11263-008-0152-6. 128

[333] A. Santos, N. McGuckin, H. Y. Nakamoto, D. Gray, S. Liss et al., Summary of travel trends: 2009 national household travel survey, United States, Federal Highway Administration, *Tech. Rep.*, 2011. 133

[334] S. Liu, *Engineering Autonomous Vehicles and Robots: The DragonFly Modular-Based Approach*, John Wiley & Sons, 2020. DOI: 10.1002/9781119570516. 133

[335] L. Liu, J. Tang, S. Liu, B. Yu, J.-L. Gaudiot, and Y. Xie, π-RT: A runtime framework to enable energy-efficient real-time robotic vision applications on heterogeneous architectures, *Computer*, 54, 2021. DOI: 10.1109/mc.2020.3015950. 133

[336] W. Fang, Y. Zhang, B. Yu, and S. Liu, Dragonfly+: FPGA-based quad-camera visual slam system for autonomous vehicles, *Proc. IEEE HotChips*, page 1, 2018. 134

[337] Q. Liu, S. Qin, B. Yu, J. Tang, and S. Liu, π-BA: Bundle adjustment hardware accelerator based on distribution of 3D-point observations, *IEEE Transactions on Computers*, 2020. DOI: 10.1109/tc.2020.2984611. 134

[338] J. Tang, B. Yu, S. Liu, Z. Zhang, W. Fang, and Y. Zhang, π-SoC: Heterogeneous SoC architecture for visual inertial SLAM applications, *IEEE/RSJ International Conference on Intelligent Robots and Systems (IROS)*, pages 8302–8307, 2018. DOI: 10.1109/iros.2018.8594181. 134, 141

[339] B. Yu, W. Hu, L. Xu, J. Tang, S. Liu, and Y. Zhu, Building the computing system for autonomous micromobility vehicles: Design constraints and architectural optimizations, *53rd Annual IEEE/ACM International Symposium on Microarchitecture (MICRO)*, pages 1067–1081, 2020. DOI: 10.1109/micro50266.2020.00089. 134

[340] Y. Gan, B. Yu, B. Tian, L. Xu, W. Hu, J. Tang, S. Liu, and Y. Zhu, Eudoxus: Characterizing and accelerating localization in autonomous machines, *IEEE International Symposium on High Performance Computer Architecture (HPCA)*, 2021. DOI: 10.1109/hpca51647.2021.00074. 134

[341] Z. Wan, Y. Zhang, A. Raychowdhury, B. Yu, Y. Zhang, and S. Liu, An energy-efficient quad-camera visual system for autonomous machines on FPGA platform, *ArXiv Preprint ArXiv:2104.00192*, 2021. 134

[342] National Highway Traffic Safety Administration, Preliminary statement of policy concerning automated vehicles, national highway traffic safety administration and others, pages 1–14, Washington, DC, 2013. 134

[343] LiDAR Specification Comparison. https://autonomoustuff.com/wp-content/uploads/2018/04/LiDAR_Comparison.pdf 137

[344] E. Kim, J. Lee, and K. G. Shin, Real-time prediction of battery power requirements for electric vehicles, *ACM/IEEE International Conference on Cyber-Physical Systems (ICCPS)*, pages 11–20, 2013. DOI: 10.1145/2502524.2502527. 137

[345] L. A. Barroso, U. Hölzle, and P. Ranganathan, The datacenter as a computer: Designing warehouse-scale machines, *Synthesis Lectures on Computer Architecture*, 13(3):i–189, 2018. DOI: 10.2200/s00874ed3v01y201809cac046. 138

[346] What it really costs to turn a car into a self-driving vehicle. https://qz.com/924212/what-it-really-costs-to-turn-a-car-into-a-self-driving-vehicle/ 138

[347] J. Jiao, Machine learning assisted high-definition map creation, *IEEE 42nd Annual Computer Software and Applications Conference (COMPSAC)*, 1:367–373, 2018. DOI: 10.1109/compsac.2018.00058. 138

[348] S. A. Mohamed, M.-H. Haghbayan, T. Westerlund, J. Heikkonen, H. Tenhunen, and J. Plosila, A survey on odometry for autonomous navigation systems, *IEEE Access*, 7:97 466–97 486, 2019. DOI: 10.1109/access.2019.2929133. 139

[349] M. Bloesch, S. Omari, M. Hutter, and R. Siegwart, Robust visual inertial odometry using a direct EKF-based approach, *IEEE/RSJ International Conference on Intelligent Robots and Systems (IROS)*, pages 298–304, 2015. DOI: 10.1109/iros.2015.7353389. 139

[350] D. Scharstein, R. Szeliski, and R. Zabih, A taxonomy and evaluation of dense two-frame stereo correspondence algorithms, *Proc. of 1st IEEE Workshop on Stereo and Multi-Baseline Vision*, 2001. DOI: 10.1109/smbv.2001.988771. 139

[351] M. Z. Brown, D. Burschka, and G. D. Hager, Advances in Computational Stereo, 2003. DOI: 10.1109/tpami.2003.1217603. 139

[352] A. Geiger, M. Roser, and R. Urtasun, Efficient large-scale stereo matching, *Proc. of the 10th Asian Conference on Computer Vision*, 2010. DOI: 10.1007/978-3-642-19315-6_3. 139

[353] J. F. Henriques, R. Caseiro, P. Martins, and J. Batista, High-speed tracking with kernelized correlation filters, *IEEE Transactions on Pattern Analysis and Machine Intelligence*, 37(3):583–596, 2014. DOI: 10.1109/tpami.2014.2345390. 139, 140

[354] J. Redmon, S. Divvala, R. Girshick, and A. Farhadi, You only look once: Unified, real-time object detection, *Proc. of the IEEE Conference on Computer Vision and Pattern Recognition*, pages 779–788, 2016. DOI: 10.1109/cvpr.2016.91. 139

[355] K. He, G. Gkioxari, P. Dollár, and R. Girshick, Mask R-CNN, *Proc. of the IEEE International Conference on Computer Vision*, pages 2961–2969, 2017. DOI: 10.1109/iccv.2017.322. 139

[356] Snapdragon Mobile Platform. https://www.qualcomm.com/snapdragon 140

[357] NVIDIA Tegra. https://www.nvidia.com/object/tegra-features.html 140

[358] G. Loianno, C. Brunner, G. McGrath, and V. Kumar, Estimation, control, and planning for aggressive flight with a small quadrotor with a single camera and IMU, *IEEE Robotics and Automation Letters*, 2(2):404–411, 2016. DOI: 10.1109/lra.2016.2633290. 140

[359] L. Liu, S. Liu, Z. Zhang, B. Yu, J. Tang, and Y. Xie, PIRT: A runtime framework to enable energy-efficient real-time robotic applications on heterogeneous architectures, *ArXiv Preprint ArXiv:1802.08359*, 2018. 140

[360] N. de Palézieux, T. Nägeli, and O. Hilliges, Duo-VIO: Fast, light-weight, stereo inertial odometry, *IEEE/RSJ International Conference on Intelligent Robots and Systems (IROS)*, pages 2237–2242, 2016. DOI: 10.1109/iros.2016.7759350. 140

[361] NVIDIA Jetson TX2 Module. https://www.nvidia.com/en-us/autonomous-machines/embedded-systems-dev-kits-modules/ 140

[362] P. Yedlapalli, N. C. Nachiappan, N. Soundararajan, A. Sivasubramaniam, M. T. Kandemir, and C. R. Das, Short-circuiting memory traffic in handheld platforms, *47th Annual IEEE/ACM International Symposium on Microarchitecture*, pages 166–177, 2014. DOI: 10.1109/micro.2014.60. 141

[363] N. C. Nachiappan, H. Zhang, J. Ryoo, N. Soundararajan, A. Sivasubramaniam, M. T. Kandemir, R. Iyer, and C. R. Das, VIP: Virtualizing IP chains on handheld platforms, *Proc. of ISCA*, 2015. DOI: 10.1145/2749469.2750382. 141

[364] NXP ADAS and Highly Automated Driving Solutions for Automotive. https://www.nxp.com/applications/solutions/automotive/adas-and-highly-automated-driving:ADAS--AND-AUTONOMOUS-DRIVING 141

[365] Intel MobilEye Autonomous Driving Solutions. https://www.mobileye.com/ 141

[366] Nvidia AI. https://www.nvidia.com/en-us/deep-learning-ai/ 141

[367] MIPI Camera Serial Interface 2 (MIPI CSI-2). https://www.mipi.org/specifications/csi-2 143

[368] Regulus ISP. https://www.xilinx.com/products/intellectual-property/1-r6gz2f.html 143

[369] J. Sacks, D. Mahajan, R. C. Lawson, and H. Esmaeilzadeh, Robox: An end-to-end solution to accelerate autonomous control in robotics, *Proc. of the 45th Annual International Symposium on Computer Architecture*, pages 479–490, IEEE Press, 2018. DOI: 10.1109/isca.2018.00047. 147, 148

[370] H. Fan, F. Zhu, C. Liu, L. Zhang, L. Zhuang, D. Li, W. Zhu, J. Hu, H. Li, and Q. Kong, Baidu apollo EM motion planner, *ArXiv Preprint ArXiv:1807.08048*, 2018. 147

[371] A. Suleiman, Z. Zhang, L. Carlone, S. Karaman, and V. Sze, Navion: A fully integrated energy-efficient visual-inertial odometry accelerator for autonomous navigation of nano drones, *IEEE Symposium on VLSI Circuits*, pages 133–134, 2018. DOI: 10.1109/vlsic.2018.8502279. 148

[372] J. Zhang and S. Singh, Visual-lidar odometry and mapping: Low-drift, robust, and fast, *IEEE International Conference on Robotics and Automation (ICRA)*, pages 2174–2181, 2015. DOI: 10.1109/icra.2015.7139486. 148

[373] N. P. Jouppi, C. Young, N. Patil, D. Patterson, G. Agrawal, R. Bajwa, S. Bates, S. Bhatia, N. Boden, and A. Borchers, In-datacenter performance analysis of a tensor processing unit, *ArXiv Preprint ArXiv:1704.04760 (published in ACM ISCA)*, 2017. 148

[374] Y.-H. Chen, J. Emer, and V. Sze, Eyeriss: A spatial architecture for energy-efficient dataflow for convolutional neural networks, *ACM SIGARCH Computer Architecture News*, 44(3):367–379, IEEE Press, 2016. DOI: 10.1145/3007787.3001177. 148

[375] Y. Chen, T. Luo, S. Liu, S. Zhang, L. He, J. Wang, L. Li, T. Chen, Z. Xu, N. Sun, and O. Temam, Dadiannao: A machine-learning supercomputer, *Proc. of the 47th Annual IEEE/ACM International Symposium on Microarchitecture, IEEE Computer Society*, pages 609–622, 2014. DOI: 10.1109/micro.2014.58. 148

[376] A. Parashar, M. Rhu, A. Mukkara, A. Puglielli, R. Venkatesan, B. Khailany, J. Emer, S. W. Keckler, and W. J. Dally, SCNN: An accelerator for compressed-sparse convolutional neural networks, *Proc. of the 44th Annual International Symposium on Computer Architecture, ACM*, pages 27–40, 2017. DOI: 10.1145/3079856.3080254. 148

[377] P. Chi, S. Li, C. Xu, T. Zhang, J. Zhao, Y. Liu, Y. Wang, and Y. Xie, Prime: A novel processing-in-memory architecture for neural network computation in ReRAM-based main memory, *ACM SIGARCH Computer Architecture News*, 44(3):27–39, IEEE Press, 2016. DOI: 10.1145/3007787.3001140. 148

[378] D. Wofk, F. Ma, T.-J. Yang, S. Karaman, and V. Sze, FastDepth: Fast monocular depth estimation on embedded systems, *ArXiv Preprint ArXiv:1903.03273*, 2019. DOI: 10.1109/icra.2019.8794182. 148

[379] M. D. Hill and V. J. Reddi, Accelerator level parallelism, *ArXiv Preprint ArXiv:1907.02064*, 2019. 148

[380] M. Hill and V. J. Reddi, Gables: A roofline model for mobile SoCs, *IEEE International Symposium on High Performance Computer Architecture (HPCA)*, pages 317–330, 2019. DOI: 10.1109/hpca.2019.00047. 148

[381] P. L. Mckerracher, R. P. Cain, J. C. Barnett, W. S. Green, and J. D. Kinnison, Design and test of field programmable gate arrays in space applications, 1992. 149

[382] R. Gaillard, Single event effects: Mechanisms and classification, *Soft Errors in Modern Electronic Systems*, pages 27–54, Springer, 2011. DOI: 10.1007/978-1-4419-6993-4_2. 149

[383] M. Wirthlin, FPGAs operating in a radiation environment: Lessons learned from FPGAs in space, *Journal of Instrumentation*, 8(2):C02020, 2013. DOI: 10.1088/1748-0221/8/02/c02020. 150

[384] F. Brosser and E. Milh, SEU mitigation techniques for advanced reprogrammable FPGA in space, Master's thesis, 2014. 150, 151

[385] B. Ahmed and C. Basha, Fault mitigation strategies for reliable FPGA architectures, Ph.D. Dissertation, Rennes 1, 2016. 150, 151

[386] S. Habinc, Suitability of reprogrammable FPGAs in space applications, *Gaisler Research, Feasibility Report*, 2002. 150

[387] G. Lentaris, K. Maragos, I. Stratakos, L. Papadopoulos, O. Papanikolaou, D. Soudris, M. Lourakis, X. Zabulis, D. Gonzalez-Arjona, and G. Furano, High-performance embedded computing in space: Evaluation of platforms for vision-based navigation, *Journal of Aerospace Information Systems*, 15(4):178–192, 2018. DOI: 10.2514/1.i010555. 151

[388] G. Lentaris, I. Stamoulias, D. Diamantopoulos, K. Maragos, K. Siozios, D. Soudris, M. A. Rodrigalvarez, M. Lourakis, X. Zabulis, I. Kostavelis et al., Spartan/sextant/compass: Advancing space rover vision via reconfigurable platforms, *International Symposium on Applied Reconfigurable Computing*, pages 475–486, Springer, 2015. DOI: 10.1007/978-3-319-16214-0_44. 151

[389] C. G. Harris, M. Stephens et al., A combined corner and edge detector, *Alvey Vision Conference*, 15(50):10–5244, Citeseer, 1988. DOI: 10.5244/c.2.23. 153

[390] E. Rosten, R. Porter, and T. Drummond, Faster and better: A machine learning approach to corner detection, *IEEE Transactions on Pattern Analysis and Machine Intelligence*, 32(1):105–119, 2008. DOI: 10.1109/tpami.2008.275. 153

[391] D. G. Lowe, Distinctive image features from scale-invariant keypoints, *International Journal of Computer Vision*, 60(2):91–110, 2004. DOI: 10.1023/b:visi.0000029664.99615.94. 153

[392] H. Bay, T. Tuytelaars, and L. Van Gool, Surf: Speeded up robust features, *European Conference on Computer Vision*, pages 404–417, Springer, 2006. DOI: 10.1007/11744023_32. 153

[393] G. Lentaris, I. Stamoulias, D. Soudris, and M. Lourakis, HW/SW codesign and FPGA acceleration of visual odometry algorithms for rover navigation on Mars, *IEEE Transactions on Circuits and Systems for Video Technology*, 26(8):1563–1577, 2015. DOI: 10.1109/tcsvt.2015.2452781. 153

[394] T. M. Howard, A. Morfopoulos, J. Morrison, Y. Kuwata, C. Villalpando, L. Matthies, and M. McHenry, Enabling continuous planetary rover navigation through FPGA stereo and visual odometry, *IEEE Aerospace Conference*, pages 1–9, 2012. DOI: 10.1109/aero.2012.6187041. 153

[395] R. S. Sutton and A. G. Barto, *Reinforcement Learning: An Introduction*, MIT Press, 2018. DOI: 10.1109/tnn.1998.712192. 153

[396] P. R. Gankidi and J. Thangavelautham, FPGA architecture for deep learning and its application to planetary robotics, *IEEE Aerospace Conference*, pages 1–9, 2017. DOI: 10.1109/aero.2017.7943929. 154

[397] T. Y. Li and S. Liu, Enabling commercial autonomous robotic space explorers, *IEEE Potentials*, 39(1):29–36, 2019. DOI: 10.1109/mpot.2019.2935338. 154

[398] D. Ratter, FPGAs on Mars, *Xcell Journal*, 50:8–11, 2004. 155, 156

[399] J. F. Bell III, S. Squyres, K. E. Herkenhoff, J. Maki, H. Arneson, D. Brown, S. Collins, A. Dingizian, S. Elliot, E. Hagerott et al., Mars exploration rover Athena panoramic camera (pancam) investigation, *Journal of Geophysical Research: Planets*, 108(E12), 2003. DOI: 10.1029/2003je002070. 155, 156

[400] Space flight system design and environmental test, 2020. https://www.nasa.gov/sites/default/files/atoms/files/std8070.1.pdf 156

[401] M. C. Malin, M. A. Ravine, M. A. Caplinger, F. Tony Ghaemi, J. A. Schaffner, J. N. Maki, J. F. Bell III, J. F. Cameron, W. E. Dietrich, K. S. Edgett et al., The Mars science laboratory (MSL) mast cameras and descent imager: Investigation and instrument descriptions, *Earth and Space Science*, 4(8):506–539, 2017. DOI: 10.1002/2016ea000252. 156, 157

[402] C. D. Edwards, T. C. Jedrey, A. Devereaux, R. DePaula, and M. Dapore, The electra proximity link payload for Mars relay telecommunications and navigation, 2003. DOI: 10.2514/6.iac-03-q.3.a.06. 157

[403] A. Johnson, S. Aaron, J. Chang, Y. Cheng, J. Montgomery, S. Mohan, S. Schroeder, B. Tweddle, N. Trawny, and J. Zheng, The lander vision system for Mars 2020 entry descent and landing, *NASA, Jet Propulsion Laboratory*, California Institute of Technology, 2017. 157

[404] Y.-H. Lai, Y. Chi, Y. Hu, J. Wang, C. H. Yu, Y. Zhou, J. Cong, and Z. Zhang, HeteroCL: A multi-paradigm programming infrastructure for software-defined reconfigurable computing, *Proc. of the ACM/SIGDA International Symposium on Field-Programmable Gate Arrays*, pages 242–251, 2019. DOI: 10.1145/3289602.3293910. 161

[405] S. Liu, J.-L. Gaudiot, and H. Kasahara, Engineering education in the age of autonomous machines, *Computer*, 54, 2021. DOI: 10.1109/mc.2021.3057407. 161

Authors' Biographies

SHAOSHAN LIU

Dr. Shaoshan Liu is founder and CEO of PerceptIn Inc., a company focusing on developing autonomous driving technologies. Dr. Liu has published more than 70 research papers, 40 U.S. patents, and over 150 international patents on autonomous driving technologies and robotics, as well as 2 books on autonomous driving technologies: *Creating Autonomous Vehicle Systems* (Morgan & Claypool) and *Engineering Autonomous Vehicles and Robots: The DragonFly Modular–Based Approach* (Wiley–IEEE). He is a senior member of IEEE, a Distinguished Speaker of the IEEE Computer Society, a Distinguished Speaker of ACM, and a founder of the IEEE Special Technical Community on Autonomous Driving Technologies. Dr. Liu received a Master's of Public Administration (MPA) from Harvard Kennedy School and a Ph.D. in Computer Engineering from University of California, Irvine.

ZISHEN WAN

Zishen Wan received his M.S. degree from Harvard University, Cambridge, MA, USA, in 2020 and his B.S. degree from Harbin Institute of Technology, Harbin, China, in 2018, both in electrical engineering. He is currently pursuing a Ph.D. in Electrical and Computer Engineering from Georgia Institute of Technology, Atlanta, GA, USA. He has a broad research interest in VLSI design, computer architecture, machine learning, and edge intelligence, with a focus on energy-efficiency, robust hardware, and system design for autonomous machines. He has received the Best Paper Award in DAC 2020 and CAL 2020.

BO YU

Dr. Bo Yu received his B.S. degree in Electronic Technology and Science from Tianjin University, Tianjin, China, in 2006, and a Ph.D. degree from the Institute of Microelectronics, Tsinghua University, Beijing, China, in 2013. He is currently the CTO of PerceptIn Inc., a company focusing on providing visual perception solutions for robotics and autonomous driving. His current research interests include algorithm and systems for robotics and autonomous vehicles. Dr. Yu is also a Founding Member of the IEEE Special Technical Community on Autonomous Driving and a senior member of IEEE.

YU WANG

Dr. Yu Wang received his B.S. degree in 2002 and Ph.D. (with honor) in 2007 from Tsinghua University, Beijing, China. He is currently a Tenured Professor and Chair with the Department of Electronic Engineering, Tsinghua University. His research interests include application-specific hardware computing, parallel circuit analysis, and power/reliability aware system design methodology. Dr. Wang has authored and coauthored over 250 papers in refereed journals and conferences. He has received Best Paper Award in ASPDAC 2019, FPGA 2017, NVMSA17, ISVLSI 2012, and Best Poster Award in HEART 2012 with 9 Best Paper Nominations. He is a recipient of DAC Under-40 Innovator Award in 2018. He served as TPC chair for ICFPT 2019, ISVLSI 2018, ICFPT 2011 and Finance Chair of ISLPED 2012-2016, and served as the program committee member for leading conferences in EDA/FPGA area. Currently he serves as Associate Editor for *IEEE Transasctions on CAS for Video Technology*, *IEEE Transactions on CAD*, and *ACM TECS*. He is an IEEE/ACM senior member. He is the co-founder of Deephi Tech (acquired by Xilinx in 2018), which is a leading deep learning computing platform provider.

Printed in the United States
by Baker & Taylor Publisher Services